READER'S DIGEST
DIY
WIRING & LIGHTING

READER'S DIGEST

DIY

WIRING &
LIGHTING

STEP BY STEP INSTRUCTIONS • EXPERT GUIDANCE • HELPFUL TIPS

Published by
The Reader's Digest Association, Inc.
London • New York • Sydney • Montreal

Contents

Introduction

Understanding the system

Maintenance and repairs

Wiring

Fixed appliances

Wiring regulations

All new domestic wiring work in England and Wales must now comply with *Part P* of the Building Regulations. *Part P: Electrical Safety* covers the design and installation of electrical wiring within the home. Ways of meeting the requirements of *Part P* are detailed in the *Approved Document*, which can be downloaded from the Government Planning Portal website. All electrical installation work in the home must also comply with the latest (17th) edition of the IEE Wiring Regulations (BS7671:2008, incorporating Amendment 1), produced by the Institution of Engineering and Technology (IET).

What you need to know:

1 **You can still do your own wiring work, but it must comply with the Wiring Regulations.**

2 **For most repairs, replacements and simple additions to wiring, you do not need to notify your local authority.**

3 **You must notify your local authority building control department before you start all other wiring jobs and pay a fee for inspection and testing.**

🚫 BUILDING REGULATIONS

All projects in this book that carry this symbol must be notified to your local authority Building Control department. Contact them if you are in any doubt about whether a job you intend to carry out needs notification.

Getting approval

DIY electrical work is still permitted following the introduction of *Part P*. What has changed is that for certain jobs (see 'Notifiable work', right), you must tell your local authority building control department what you intend to do before you start the work. As with all other building work covered by the Building Regulations, this is done either by a *Building Notice* or by submitting *Full Plans*. A fee, which includes inspection and testing, is payable and – with a Full Plans application – a completion certificate is issued when the work is finished satisfactorily.

If you decide to use professional electricians to do the work, they will be able to self-certify that the work complies with the Building Regulations, provided they are registered as 'competent persons' for *Part P*.

Non-Part P registered electricians should be able to issue a BS7671 installation certificate, but the work still needs to be notified to the local authority.

Non-notifiable work

Some minor DIY wiring jobs do not need to be notified to the local authority. These include:

- **Replacing existing wiring accessories, such as switches, socket outlets and ceiling roses.**
- **Replacing their mounting boxes.**
- **Replacing a single circuit cable that has been damaged by, for example, impact, fire or rodent attack.**
- **Installing or upgrading main or supplementary equipotential bonding (see page 32)**
- **Adding new lighting fittings and switches to an existing lighting circuit, unless the work is in a kitchen or bathroom or outdoors (see top right).**
- **Adding new socket outlets or fused spurs to an existing ring main or radial circuit, unless the work is in a kitchen or bathroom or outdoors (see top right).**

If you do your own non-notifiable wiring, you might want to employ an electrician to test and inspect it to make sure it is safe, but you cannot use an electrician to certify your own notifiable work.

Notifiable work

Certain wiring jobs must be notified to your local building control department before you start work. These include:

- **The installation of any new indoor circuit, such as one supplying an electric shower or a home extension.**
- **The installation of any new outdoor circuit, such as one supplying garden lighting or an outbuilding.**
- **Consumer unit replacement.**
- **Installation of underfloor heating.**
- **Any new wiring work in kitchens or bathrooms, including the extension of the existing circuit, through straightforward replacement is not notifiable.**
- **The installation of extra-low-voltage lighting circuits. But the installation on existing lighting circuits of pre-assembled extra-low-voltage light fittings with CE approval is exempt.**

If you are in doubt as to whether certain work requires notification, contact your local authority building control department.

Employing an electrician

One of the reasons Part P was introduced was to prevent non-qualified tradesmen from carrying out electrical work. Employing a Part P registered electrician to carry out electrical work in the home has the advantage that they can self-certify their work without the need to notify the local authority building control department.

Qualified electricians are members of the Electrical Contractors' Association (ECA) or on the roll of the National Inspection Council for Electrical Installation Contracting (NICEIC) – see *Useful Contacts* page 11.

New cable core colours

Since 1 April 2006, new fixed wiring work in the home will have been done with cable using new colours for the insulation of the main cores. Cable with the old core colours is now not available to buy. This change has been introduced as part of a process of product harmonisation across the European Union, many years after flex core colours were changed from red and black to brown and blue. See page 56 for more details.

Where you have a mixture of old and new wiring from altering or extending an existing wiring installation, you must place a warning notice at the fuse board or consumer unit. Its wording is as follows:

Caution

This installation has wiring colours to two versions of **BS7671**. Great care should be taken before undertaking extension, alteration or repair that all conductors are correctly identified.

The illustrations in this book show new wiring being carried out using cable with new core colours. When using two-core-and-earth cable with the new core colours, remember that new brown = old red and new blue = old black. See page 56 if you are using three-core-and-earth cable.

NEW WIRING REGULATIONS

The 17th Edition of the Wiring Regulations, which came into effect in 2008, introduced some radical changes for domestic wiring:

- **All new cable that you bury or conceal at a depth of less than 50mm deep in walls must either be stringently mechanically protected or be protected by an RCD (residual-current device) that switches off electricity automatically if there is a fault.**
- **All new socket outlets must be RCD protected – either by an RCD in the socket itself or by having an RCD in the circuit to which it has been added.**
- **All new wiring for bathrooms must be protected by an RCD (as must all changes to existing bathroom wiring).**

These changes apply only to new wiring work – not to existing installations.

Regulations in Scotland and Northern Ireland may differ. For Northern Ireland, check with the Building Control department at your local council; for Scotland, the Building Standards Services of your local council.

United Kingdom rules do not apply in the Republic of Ireland where electrical installations must comply with Ireland's National Rules for Electrical Installations. For information contact your local government of the Electro-Technical Council for Ireland: www.etci.ie

Electrical emergencies

Warning: the main on-off switch on your consumer unit disconnects only the fuses or miniature circuit breakers (MCBs) and the cables leading out from it to the household circuits. It does NOT disconnect the cables entering via the meter from the service cable. Do not tamper with these cables. They are always live at mains voltage.

Fire in an appliance

1 If a plug-in appliance is on fire, switch the appliance off at the socket outlet and pull out the plug.

2 If a fixed appliance with no plug is on fire, turn it off at the wall switch if you can, or at the main switch on the consumer unit (see below).

3 Do not use water on an electrical fire. Smother the fire with a rug or blanket, or use a dry-powder fire extinguisher.

4 Get the appliance checked (and repaired if possible) by an expert before you use it again.

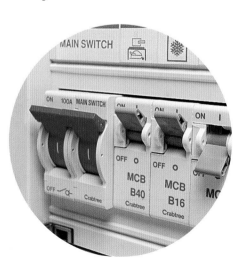

Smell of overheating

1 If you smell burning from an appliance, turn off the switch at the socket and pull out the plug. If it is a fixed appliance with no plug, turn off its wall switch or the main switch at the consumer unit. Turn off the appliance switch. Check flex connections and re-connect if necessary; if they are sound, have the appliance checked by an expert.

2 If the smell comes from a socket outlet or a plug, turn off the main switch at the consumer unit. If the plug is hot, let it cool before unplugging it. Then check its connections, including the fuse contacts, and examine the flex for damage. Replace as necessary (page 38). If the socket is hot, check it for faulty connections and renew as necessary (page 50).

No electricity

1 If power throughout your house fails and neighbouring houses are also without power, there is a mains supply failure. Report it to the 24-hour emergency number under 'Electricity' in the phone book.

2 If your system is protected by a whole-house residual current device (RCD), check whether it has switched itself off. Try to switch it on again if it has.

3 If it will not switch on, the fault that tripped it off is still present on the system. **Call an electrician to track it down and rectify it.**

4 If you do not have an RCD and your house is the only one without power, there may be a fault in your supply cable or your main supply fuse may have blown. Do not touch it. Report the power failure as described above.

Minor emergencies

1 If one appliance fails to work, unplug it and check its plug, fuse and flex; renew them as necessary. If the appliance still fails to work, plug it in a different socket outlet to test it. If it works, the problem is with the original socket; if not, take the appliance to an expert for repair.

2 If all lights or appliances on one circuit stop working, switch off at the consumer unit and check the circuit fuse or MCB (page 36). If it is sound, there may be a fault in the circuit cable. **Call in an electrician to rectify it.**

Electric shock

Warning: If you get a minor shock from an electrical appliance, a plug or other wiring accessory, stop using it immediately.

• Get a repair expert to check the appliance for earth safety, and replace damaged plugs and wiring accessories as soon as possible. Only use PVC insulating tape to make a temporary repair.

• If someone receives a major shock, DO NOT touch bare flesh while the person is in contact with the source of the current. If you do, the current will pass through you as well, giving you an electric shock.

1 Immediately turn off the source of the current if you can.

2 If you cannot do this, grab the person's clothing and drag them away from the source of the current, or stand on some insulating material such as a book, and use a broom or a similar wooden object to move the person or the current source.

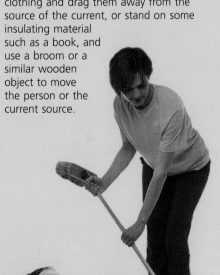

3 Lay a conscious but visibly shocked person flat on their back with their legs raised slightly and cover with a blanket. Do not give food, drink or cigarettes. Cool visible burns with cold water, then cover them with a dry sterile dressing. Do not apply ointments. Call an ambulance.

If someone is unconscious Place an unconscious person in the recovery position. Tilt the head back and bring the jaw forward to keep the airway clear. Cover them with a blanket and call an ambulance.

Check the person's breathing Monitor breathing and heartbeat continuously until the ambulance arrives. If either stops, give artificial ventilation or external chest compression as necessary, if you are trained to do so.

Electrical safety

Electricity can kill – either through receiving an electric shock or because faulty wiring has caused a fire. Follow these safety precautions to keep yourself safe when using electrical appliances or working with or near electric circuits. See pages 6–7 for details on the Building Regulations governing electrical safety in the home.

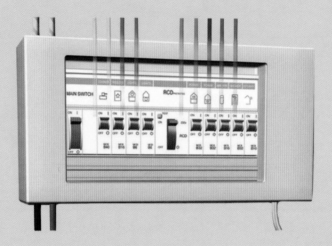

Isolate mains circuits ▲
at the consumer unit (see page 16) by switching off the miniature circuit breakers (MCBs) or by removing the circuit fuses before carrying out any work on the house wiring.

Always unplug appliances ▼
from the mains before attempting any repair work on them.

Uncoil extension leads fully
before using them. If the lead is powering any appliance with a heating element, check that the current rating is suitable for the appliance wattage – e.g. 6A for up to 1.4kW, 10A for up to 2300W and 13A for up to 3kW.

Do not overload socket outlets ▼
– either mechanically, by using adaptors, or electrically, by plugging in too many high-wattage appliances. If you are using a four or six-way adaptor, refer to the label on the back of the adaptor to check the maximum load (normally 13A for the whole adaptor).

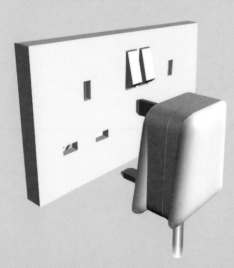

Check appliance plugs and flexes regularly for damage, cuts or other signs of wear. Replace damaged parts at the earliest possible opportunity.

Replace blown circuit fuses using fuse wire or cartridge fuses of the correct rating (see page 38). Never use any other metallic object to repair a fuse.

Out of doors ▶
plug any power tool being used into a residual current device (RCD) adaptor. It is worth considering the installation of RCDs to protect all the socket outlet circuits in your house.

Check there are earth connections for all appliances and wiring accessories, and earth all metal pipework and plumbing fittings. The only situation where an earth connection is not needed is in the flex to a non-metallic lampholder or to power tools and portable appliances that are double-insulated.

Keep water and electricity apart Never plug in appliances or operate electrical switches with wet hands. Never take an electrical appliance into the bathroom, even if the flex is long enough. Never use electrical equipment outside the house in wet conditions.

Use an electronic detector to help you avoid drilling into pipes or cables already buried or concealed in the wall and to enable you to screw directly into floor/ceiling joists or the studs of partition walls. These work in one of three ways: metal detection (pipes and cables), density detection (studs and joists) and voltage detection (live electric cables). You will find detectors with one, two or all three of the functions. Note that metal detectors may not work with foil-backed plasterboard. ▶

Useful contacts

Electrical Contractors' Association
ESCA House
34 Palace Court
London W2 4HY
020 7313 4800
www.eca.co.uk

National Inspection Council for Electrical Installation Contracting (NICEIC)
Warwick House
Houghton Hall Park
Houghton Regis
Dunstable
Bedfordshire LU5 5ZX
0870 013 0382
www.niceic.com

Institution of Engineering & Technology (IET)
Michael Faraday House
Six Hills Way
Stevenage SG1 2AY
01438 313311
www.theiet.org

Local Authority Building Control
020 7091 6860
www.labc.uk.com

Government Planning Portal
www.planningportal.gov.uk

Understanding
the system

The electrical system

Before you do any electrical work you should get to grips with the different types of circuit in your home.

Electric power is measured in **watts** (W). The flow of electricity is called current, and is measured in **amps** (A). The driving force, or pressure, of the current is measured in **volts** (V). The pressure of public supply in Britain has been standardised at 230 volts. In Britain, mains electricity is **alternating current** (AC) and the electricity from batteries is **direct current** (DC). The advantage of alternating current is that it can be transformed from one voltage to another so a power station can supply a very high voltage to substations that reduce the voltage to 230V to supply homes.

■ **Lighting circuit** The circuit runs out from the consumer unit, linking a chain of lighting points. A cable runs from each lighting point to its switch. The circuit is protected by a 5amp circuit fuse or 6amp MCB. It can safely supply up to a maximum of about 1200 watts. Overloading is not a problem with modern energy-saving light bulbs, but a home should have at least two separate lighting circuits.

■ **Ring main circuit** The circuit is wired as a ring that starts from the consumer unit and returns to it, allowing current to flow to socket outlets either way round the ring. It can serve a floor area of up to 100m². It is protected by a 30amp circuit fuse or 32amp MCB. It can have any number of sockets or fused connection units on it, but its maximum total load is about 7000 watts. For larger total loads and larger floor areas, additional ring circuits are needed.

■ **Spur on a ring circuit** Extra socket outlets can be added to an existing ring main circuit via 'spurs' branching off the ring at a socket outlet or junction box. In theory, each outlet could supply a spur to a single or double socket or a fused connection unit. However, the circuit including any spurs must not serve rooms with a floor area of more than 100m² – and its maximum load is still 7000 watts.

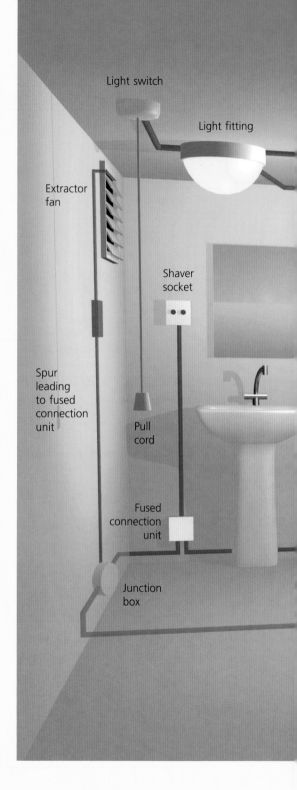

Light switch

Light fitting

Extractor fan

Shaver socket

Spur leading to fused connection unit

Pull cord

Fused connection unit

Junction box

Radial socket outlet circuit (not shown). Wired with a single cable going from socket to socket. Two circuits commonly used: 20A using 2.5mm² cable (up to 50m² floor area – e.g. extension or garage) and 30A/32A using 4 mm² cable (up to 75 m² – e.g. kitchen).

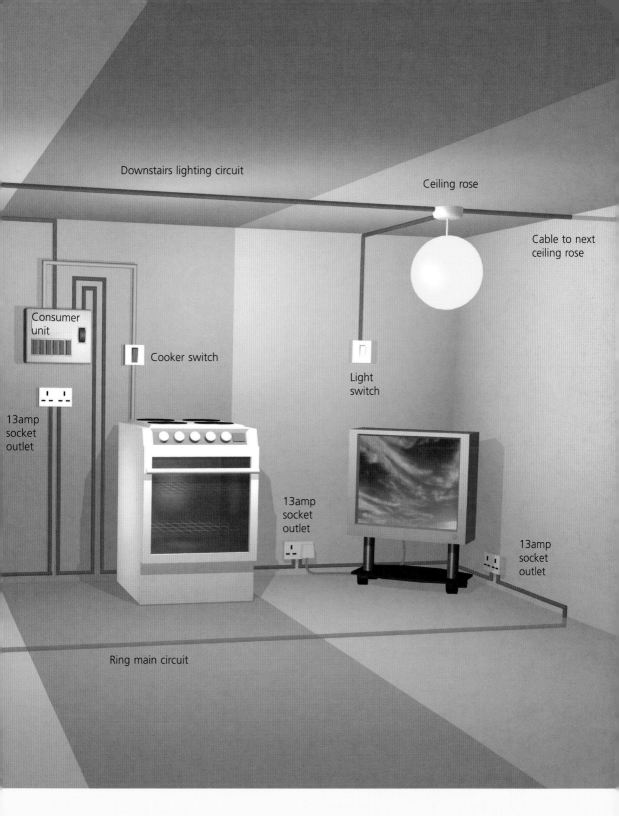

Downstairs lighting circuit

Ceiling rose

Cable to next ceiling rose

Consumer unit

Cooker switch

Light switch

13amp socket outlet

13amp socket outlet

13amp socket outlet

Ring main circuit

Single-appliance circuit An appliance that is a large consumer of electricity and in either in constant or frequent use – a cooker, a fixed water heater, or a shower heater unit, for example – has its own circuit running from the consumer unit. It would take too large a proportion of the power available on a shared circuit and would be likely to cause an overload.

A freezer may also have its own (non RCD-protected) circuit, serving just the one (labelled) socket outlet. This will protect the freezer contents in the event of an RCD trip.

The consumer unit

Modern fuse boards – called consumer units – may look different from home to home, but the basic components are the same.

Consumer unit The householder's responsibility for the system begins here. It houses the main on-off switch, the main earthing terminal block for all the house circuits, and individual fuses or miniature circuit breakers (MCBs) for each circuit. Some modern consumer units have blanked-off spaces for additional MCBs to be installed at a later date.

• The number of circuits varies according to a household's needs, but always includes separate lighting and power circuits.
• Label the MCBs to show which circuit each one protects. To identify the circuits, turn off the main switch and switch off one MCB at a time. Turn the main switch back on and check which lights or appliances are *not* working.

Service cable Electricity enters the home through the service (supply) cable – usually buried underground in urban areas, but may be run overhead in rural areas. It carries electricity at 230 volts. The current flows along the live conductor and returns along the neutral conductor. Never interfere with the service cable, which is the property of your electricity supply company. The term 'live' has been replaced by 'line' or 'phase' in the electrical industry. Live and neutral imply that current flows only in the live conductor, whereas both carry current at all times. The terms live and neutral are used throughout this book for clarity.

Miniature circuit breakers (MCBs)
Modern consumer units have MCBs instead of fuses. If too much current is demanded, the circuit is disconnected instantly and a switch moves to the 'off' position or a button pops out. Reset the switch or button to restore power to the circuit.

MCB CURRENT RATINGS

As part of a move towards European standardisation, the ratings marked on new MCBs are being changed.
  5amp becomes   6amp
15amp becomes 16amp
30amp becomes 32amp
45amp becomes 40amp (or 50amp)

Sealed unit/service cut-out The service cable ends here. Its neutral conductor is connected to a solid terminal. Its live conductor is connected to a fuse (the service cut-out), which is usually rated at 60amps or, in modern installations, at 100amps. It is a deliberate weak link that will melt and disconnect the supply to the house if more current is demanded than the service cable can safely supply.

Do not tamper with the sealed unit.

OLD FUSE BOXES

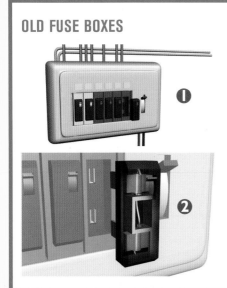

In older systems, each circuit is protected by a fuse, not an MCB. These fuseholders have slot-in rewirable fuse carriers or sealed cartridges ❶.
    Each fuseholder is marked with the rating in amps (A) of the fuse if contains. A lighting circuit is protected by a 5amp fuse, and a ring main circuit by a 30amp fuse.
    Within each of the fuse carriers or cartridges is a small length of fuse wire ❷. If the current demanded by the circuit exceeds the rating, the fuse wire melts ('blows') and the circuit is disconnected. Always keep a supply of replacement wires or cartridges in a range of appropriate amp ratings.

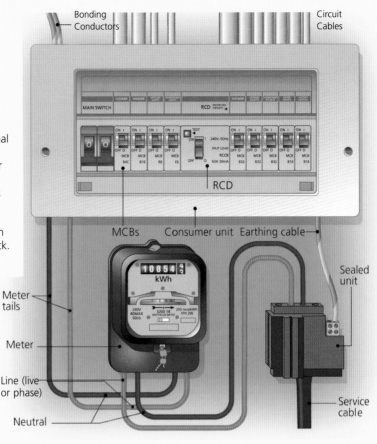

Circuit cables
Individual circuits are supplied by cables running out from the consumer unit. The live conductor in each circuit cable is connected to a terminal on its fuse or MCB. The neutral conductor connects to the main neutral terminal block in the consumer unit, and the earth conductor to the main earthing terminal block.

Residual current device (RCD)
An RCD monitors the balance of the live and neutral current flows. An imbalance occurs if current leaks from a circuit because of faulty insulation, or because someone has received an electric shock. If the RCD detects an imbalance (of just 30mA) it switches off the supply immediately.
• An existing consumer unit may have a single 30mA RCD (with a current rating of 80A or 100A) protecting all circuits, but is likely to be 'split' with RCD protection only for at-risk circuits such as those to socket outlets. This prevents a fault taking all the circuits out. An RCD in its own enclosure may have been added to an existing installation to protect at-risk circuits.
• Wiring Regulations require 30mA RCD protections on most new circuits, which can be achieved by having more than one RCD or by using RCBOs (RCDs combined with an MCB) – see page 72.

Test button

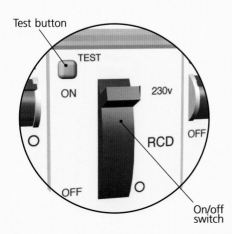

METER TAILS

Houses wired after 2006 will have brown and blue meter tails

Earthing cable This connects the main earthing terminal block in the consumer unit (to which all the circuit cables are connected) to the earthing point provided by the electricity supplier – usually on the sealed unit or the service cable. Bonding conductors connect metal gas and water supply pipework to the main earthing terminal block.

Meter A two-tariff meter with two displays may be installed to allow the use of off-peak electricity for storage heaters.

Meter tails These two cables (live and neutral) link the sealed unit to the meter and the meter to the consumer unit.
• The live cable is covered with red insulation and the neutral cable with black. Homes wired since 2006 will have a brown live cable and a blue neutral cable.
• The outer sheaths may be grey or match the colour of the insulation.
• The electricity supply company must disconnect the supply before work can be done on the meter tails.

THE CONSUMER UNIT

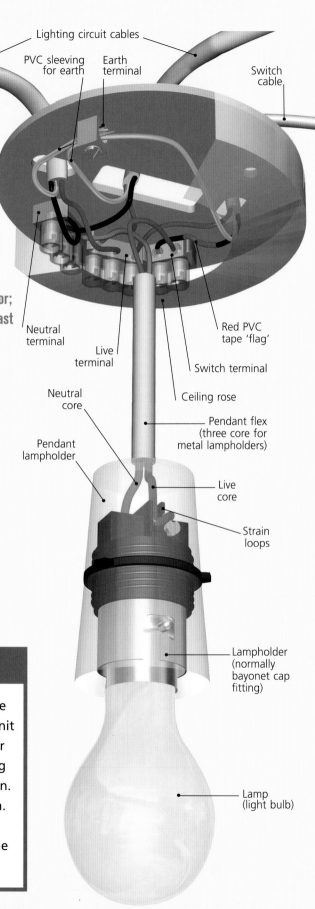

Lighting circuit cables

PVC sleeving for earth

Earth terminal

Switch cable

Neutral terminal

Live terminal

Red PVC tape 'flag'

Switch terminal

Neutral core

Ceiling rose

Pendant lampholder

Pendant flex (three core for metal lampholders)

Live core

Strain loops

Lampholder (normally bayonet cap fitting)

Lamp (light bulb)

The lighting system

All the light fittings in a circuit are connected in a chain, with switch cables connected into each lighting point. A house will usually have separate circuits for each floor; a bungalow should also have at least two lighting circuits.

Lighting cable is twin-core-and-earth, with live, neutral and earth cores. The bare earth cores are covered in slip-on green-and-yellow **PVC sleeving** when exposed. This cable carries the power for all the lights on the circuit. The flex to each fitting is of a lower rating and only needs an earth core if the lampholder is metal. Switches are connected to each light fitting. When the switch is turned on, a circuit is completed between the light and the live supply.

A flag of **red PVC insulating tape** is placed over the black core of the switch cable at the switch and at the fitting. This flag (which will be brown on a blue core for post-2006 wiring) indicates that this core is 'live' when-ever the switch is turned on.

SAFETY WARNING

Before starting work, turn the power off at the consumer unit and remove the fuseholder or switch off the MCB protecting the circuit you are working on. Turn the main switch back on. Turning off the light at the wall switch does not make the fitting safe to work on.

Lighting circuit

Each lighting circuit is wired as a radial circuit, starting at the consumer unit **①** and terminating at the last light fitting on the circuit (only first light shown).

FAULT DIAGNOSIS

SPARKING LIGHT SWITCH/ CRACKED SWITCH FACEPLATE

Worn or dirty contacts
Replace switch (see page 85).
Impact damage
Replace switch.

BROKEN CEILING ROSE COVER

Impact damage
Replace ceiling rose (see page 90).

Socket outlets

Most electrical appliances can be plugged in at any socket outlet, which may be sunk in the wall or surface mounted. Permanently fixed appliances, such as cooker hoods, are connected directly to the socket outlet circuit using fused connection units (FCUs) and are protected by the dedicated fuse in the unit.

As a plug is inserted, the **earth pin** enters first because it is longest. As the earth pin pushes in, it opens protective **shutters** in front of the holes for the **live** and **neutral pins**.

When the plug has been fully inserted, all three pins make contact with their corresponding **sprung connectors** inside the socket outlet. These are connected to the wiring terminals.

When the **switch** is moved to the ON position, the contacts complete a circuit by linking the live terminal to the live sprung connector. Power flows from the ring main via the live pin through the flex

to the appliance. The circuit is completed via the neutral pin. A **flying earth link** from the socket outlet earth terminal ensures that a metal mounting box is earthed.

FAULT DIAGNOSIS

SOCKET OUTLET NOT WORKING

Burnt or damaged outlet
Replace socket outlet (see page 50).

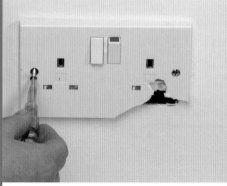

Cable disconnected inside outlet
Turn off mains and check that connections to terminals are secure.

Dirt in connectors or switch
Replace socket outlet (see page 50).

Consumer unit fuse blown or MCB tripped
Trace and rectify fault. Replace fuse or switch MCB back on (see page 16).

SAFETY CHECK

ONCE A YEAR

Test socket outlets using a socket tester

This is an inexpensive device that plugs into each socket outlet and diagnoses wiring problems.

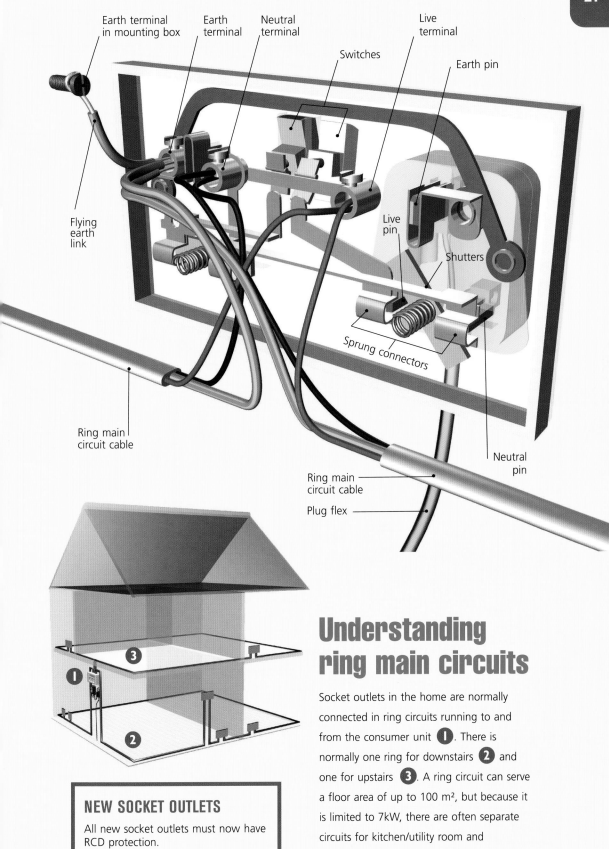

Earth terminal in mounting box

Earth terminal

Neutral terminal

Switches

Live terminal

Earth pin

Flying earth link

Live pin

Shutters

Ring main circuit cable

Sprung connectors

Neutral pin

Ring main circuit cable

Plug flex

Understanding ring main circuits

Socket outlets in the home are normally connected in ring circuits running to and from the consumer unit ❶. There is normally one ring for downstairs ❷ and one for upstairs ❸. A ring circuit can serve a floor area of up to 100 m², but because it is limited to 7kW, there are often separate circuits for kitchen/utility room and garage/workshop/garden.

NEW SOCKET OUTLETS

All new socket outlets must now have RCD protection.

Plugs and fuses

Correctly fitting the appropriate plug, flex and fuse to an electrical appliance is crucial to avoid the risks of fire and electric shocks. The key factor in deciding what rating of fuse to use is the wattage of the appliance. Modern appliances are supplied with factory-fitted moulded plugs – all that is required is to fit the correct fuse if the original blows.

The three pins of the plug are connected to the live, neutral and earth **cores** in the **flex**. The **fuse** sits between the live pin and the live flex terminal.

The earth pin is longer than the live and neutral pins so that the earth contact is made first as the plug is inserted into the socket outlet (see page 20).

On new plugs the **live** and **neutral** pins are partly insulated to prevent fingers accidentally coming into contact with the metal pins as the plug is inserted or withdrawn. Once the plug is firmly in place, and the socket outlet **switch** turned on, power can flow through the flex to the appliance.

To get power to the appliance the current must pass through the **fuse.** The fuse contains a fine strip of metal. If the current drawn from the mains is too large the fuse wire will overheat and melt, stopping the excessive current from damaging the appliance.

The earth wire forms a safety route for the electricity to flow to earth in case of electrical faults. All flex cores are individually insulated by a coating of coloured plastic, and are contained within a flexible insulating **outer sheath.**

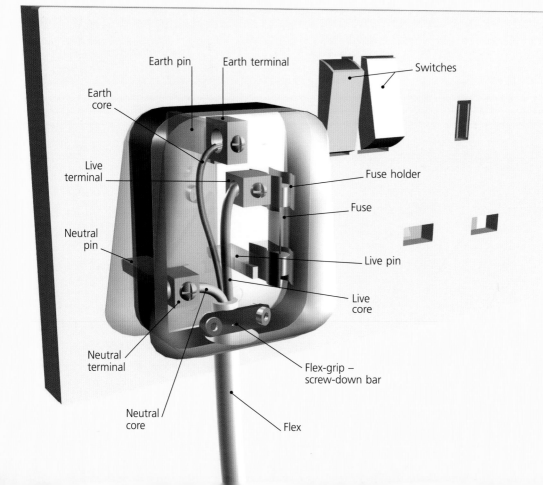

Earth pin
Earth terminal
Switches
Earth core
Live terminal
Fuse holder
Fuse
Neutral pin
Live pin
Live core
Neutral terminal
Flex-grip – screw-down bar
Neutral core
Flex

SAFETY CHECK

ONCE EVERY SIX MONTHS

Check flexes and plugs
*This is particularly important
with appliances which are
portable or which are moved
during use, such as a hair dryer,
iron, vacuum cleaner or power
tool. Replace if necessary (see
pages 39 and 44).*

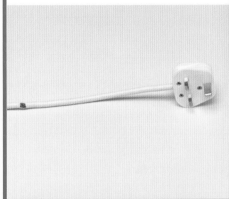

FAULT DIAGNOSIS

NO POWER TO APPLIANCE

Cracked plug/live parts exposed
Replace plug (see page 44).

Fuse blown *Unplug appliance and
check plug and flex for evidence of
scorching caused by a short circuit.
Replace flex if necessary (see page
39). Fit replacement fuse of correct
rating (see page 38) and test
the appliance.*
MCB tripped or circuit fuse
blown *Reset MCB or replace blown
fuse (see page 36). If fault reoccurs,
contact an electrician.*
Flex discontinuity *Check each core
with a continuity tester (see page
64) and replace the flex if faulty
(see page 39).*

SAFETY STANDARDS

Buy plugs marked 'Made to
British Standard BS1363', and
fuses made to BS1362. Fit
tough rubber plugs to power
tools and to garden equipment,
to prevent them being
damaged if dropped.
   If discarding a moulded-on
plug, cut through the flex
close to the plug body. Then
deform the pins with hammer
blows so the plug, if found,
cannot be inserted in a socket
outlet and cause the user to
receive a shock.

Outdoor sockets

Many electrical items, such as lawnmowers, hedge trimmers, pond pumps and garden lights are used outdoors. Water and electricity are a dangerous combination, so special switches and sockets with hinged weatherproof covers are essential for outdoor installations.

A socket outlet on an outside wall can be wired as a spur from an indoor socket outlet on the ring main circuit. Unless this circuit is already RCD protected, the new outdoor socket must have an integral RCD or be wired via a separate RCD.

The connection can be taken from a junction box or a socket outlet and PVC cable is routed through the outside wall, ideally straight into the back of the new socket mounting box. Where cable has to be run along the outside wall, it must be protected by metal or heavy-duty plastic conduit.

Drain holes in the base of the **mounting box** allow condensation build-up to escape. The spring-hinged weatherproof cover prevents rainwater from entering the terminals of the socket outlet (see page 120).

When a power supply is being run to an outbuilding, it must not be fed from a spur, but must have a dedicated RCD-protected circuit wired from the consumer unit.

For further details on outdoor wiring (and the types of cable to use), see pages 118-124.

RCD (residual current device)

Test button

On/off switch

Spur from ring main

PVC insulated cable

SAFETY WARNING

To avoid cable in the ground from being damaged by digging, outdoor cabling to a shed or outhouse should be buried at least 500mm underground. Armoured cable needs no further protection, but PVC cable must be run through protective conduit.

SAFETY CHECK

BEFORE EACH USE

Press test button on RCD to check it is operating correctly *Then press 'reset' or move switch to 'on' position to restore the power.*

FAULT DIAGNOSIS

NO FUNCTIONS

Power off *Check RCD has not tripped. Then check MCB (or fuse) serving the circuit. Reset or repair (see page 36).*

CLEANING THE SOCKET

Once a year, clean the mounting box and cover. Turn off the outdoor extension by switching the RCD to 'off'. Using a clean, damp cloth, wipe the exterior surfaces of the socket and cover. Restore the power.
   Apply a small amount of grease to each of the spring hinges (above). This will keep the action free and displace water to prevent rusting.

Weatherproof
cover

Outdoor
socket
outlet

Mounting
box

Conduit
(covering cable)

Telephone wiring

Although the master telephone socket in your home must be fitted by a telephone company, you can add extension cables and sockets yourself. The only limitation is the number of devices that can be plugged into the same phone line.

The phone signals are routed from the master socket to all the extension sockets in the house in a line.

An extension cable normally has four wires, but only three wires carry a signal to the terminals in the extension socket.

The four wires are colour coded. Blue with white rings connects to terminal two, orange with white rings to terminal three (the ringer terminal), white with orange rings to terminal four and white with blue rings to terminal five.

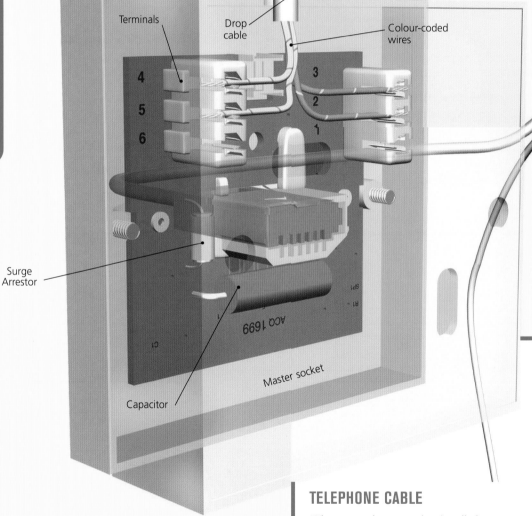

Terminals

Drop cable

Colour-coded wires

4

5

6

3

2

1

Surge Arrestor

ACQ 1699

Capacitor

Master socket

A **drop cable** carries the speech signal, a control signal and the ringing current into a **master socket** within the home.

A **capacitor** and **resistor** in the master socket separates the ring signal from the two speech signals.

TELEPHONE CABLE

When your phone number is called, a ringer signal is routed to your house from the telephone exchange. It passes through an underground or overhead cable called a drop cable ❶. The drop cable is connected to a master socket ❷, from which extensions run to other phone sockets ❸ and modem connections ❹.

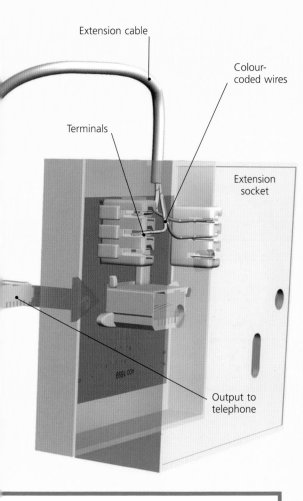

Extension cable

Colour-coded wires

Terminals

Extension socket

Output to telephone

ADD 1699

Telephone extensions

You don't need to connect into the master socket to fit a telephone extension. The kits you can buy have a 'plug' that goes into the master socket. This has its own socket for the existing phone and is attached to a length of cable which goes to the extension socket position – see page 108 for details.

Phones per line

Every telephone and modem has a rating called the Ringer Equivalence Number (REN). On most telephones this number is one (it is usually found on the underside of the phone). A single telephone circuit can handle a maximum REN of four, or four standard-rated phones. Effectively, this is the total number of sockets that the electric current in the telephone cable can supply.

SECURITY CHECK

EVERY SIX MONTHS

Check wiring *Make sure wiring fixed to skirting boards is not damaged. Also, check wiring run under doors is not crushed – re-route over doors if possible. Make sure connections into sockets are not loose, and that socket faceplates are secure.*

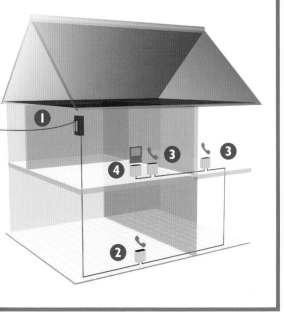

Maintenance and repairs

Finding fault with the electrical system

Your electrical system can fail at several different points. The cause may be something as simple as a blown plug fuse or faulty appliance, or there may be a fault in your circuit wiring. Follow the flow chart below to help you to diagnose the problem and work out how to fix it.

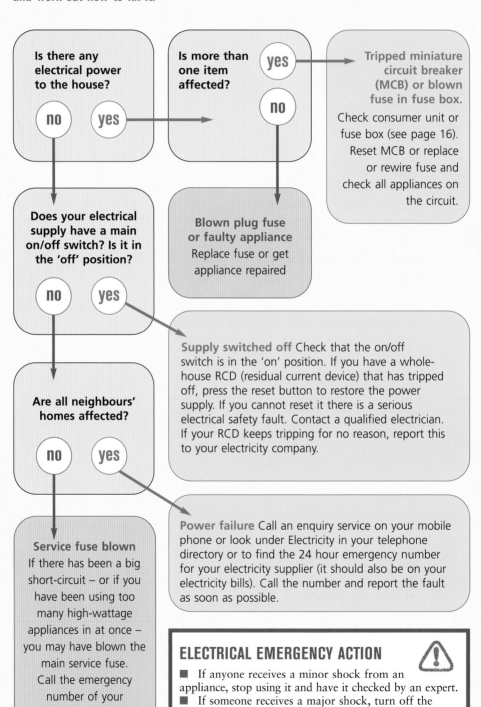

Is there any electrical power to the house?

no   yes

Is more than one item affected?

yes   no

Tripped miniature circuit breaker (MCB) or blown fuse in fuse box.
Check consumer unit or fuse box (see page 16). Reset MCB or replace or rewire fuse and check all appliances on the circuit.

Does your electrical supply have a main on/off switch? Is it in the 'off' position?

no   yes

Blown plug fuse or faulty appliance
Replace fuse or get appliance repaired

Supply switched off Check that the on/off switch is in the 'on' position. If you have a whole-house RCD (residual current device) that has tripped off, press the reset button to restore the power supply. If you cannot reset it there is a serious electrical safety fault. Contact a qualified electrician. If your RCD keeps tripping for no reason, report this to your electricity company.

Are all neighbours' homes affected?

no   yes

Service fuse blown
If there has been a big short-circuit – or if you have been using too many high-wattage appliances in at once – you may have blown the main service fuse. Call the emergency number of your electricity supplier and report the fault.

Power failure Call an enquiry service on your mobile phone or look under Electricity in your telephone directory or to find the 24 hour emergency number for your electricity supplier (it should also be on your electricity bills). Call the number and report the fault as soon as possible.

ELECTRICAL EMERGENCY ACTION

■ If anyone receives a minor shock from an appliance, stop using it and have it checked by an expert.
■ If someone receives a major shock, turn off the current immediately at the consumer unit or fuse box. Refer to the guidelines on pages 10-11.

FAULT DIAGNOSIS

MILD SHOCK FROM METAL-CASED DEVICE

Earth fault within item *Contact a qualified repairer. This fault should have tripped an RCD (see page 17), so test yours, or have one fitted.*

NO POWER

MCB tripped *Turn off appliances, reset switch and identify faulty device by unplugging all appliances, then reconnecting them one by one.*

RCD tripped *Reset RCD. If reset is impossible, call a qualified electrician.*

Local power failure *Check neighbouring buildings to verify, and report power failure to your electricity company.*

Service fuse blown *Call electricity company immediately to replace.*

MCB IN CONSUMER UNIT TRIPS REGULARLY

Circuit overload *Caused by plugging in too many high-wattage appliances at once. Make sure that you don't overload a ring circuit – maximum wattage is 7kW.*

Faulty device on circuit *If an MCB trips when a specific appliance is used, unplug it and have it checked. If an MCB trips periodically, plug in one appliance at a time to find the faulty appliance. Then have it repaired.*

A safe earthing system

The jobs described on the following pages are suitable only for modern wiring systems. Proper earthing of metal parts you can touch on electrical equipment and appliances provides vital protection from the risk of electric shock.

The earth conductor in each circuit cable is connected to the main earth terminal block in the consumer unit. This is connected in turn to an earthing point provided by the electricity supply company.

Modern installations have protective multiple earthing (PME). This is earthed via the electricity supply cable's neutral connector to the 'star point' at the local electricity supply transformer, which is connected to a permanent earth at the substation.

Non-electrical metal fittings such as plumbing pipework that might come into accidental contact with the electrical system also need earth 'bonding' (see page 32).

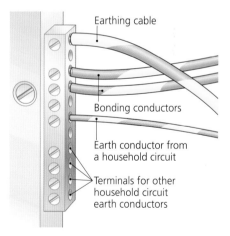

Earthing cable

Bonding conductors

Earth conductor from a household circuit

Terminals for other household circuit earth conductors

The main earth terminal block in the consumer unit has terminals for the earth conductors in all the house circuit cables. One terminal is for the main earth cable running to the main house earthing point. Other terminals take the main bonding conductors.

A house in a rural area with an overhead power supply may have no earth connection via the supply cable. The house is earthed by an earth electrode (a metal spike) driven deep into the ground close to the house. This 'TT' earthing needs additional protection, provided by a residual current device (RCD).

The house wiring is connected to the earthing system via the earth conductors (cores) in the circuit cables. The casing of electrical equipment is either earthed or insulated by the manufacturer. Check that any metal items that you install yourself, or have installed for you, are safely earthed.

Bonding metal pipes and fittings

Certain non-electrical metal items – like pipes and bathroom radiators – must be linked to the earthing system. This is known as 'bonding' and is done with single-core earth cable.

Main bonding

One or more green-and-yellow insulated cables called main bonding conductors should run from the main earth terminal block in the consumer unit to incoming metal water, gas and oil service pipes (but not to telephone or television cables). The size of the cable should be 10mm^2 or more.

Supplementary bonding

In rooms containing a bath or a shower, most homes will have supplementary 'equipotential' bonding where all metal items – such as radiators, towel rails, supply pipes and metal baths – are linked together by earth cables and are also linked to the earth conductors in circuits supplying the bathroom (for a shower or heated towel rail, for example). This prevents you getting an electrical shock if you were to touch one of these items while in contact with faulty electrical equipment.

SUPPLEMENTARY BONDING

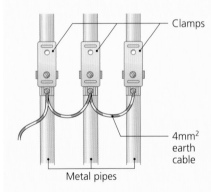

Link airing cupboard pipes to ensure the pipework offers an unbroken metal earth route.

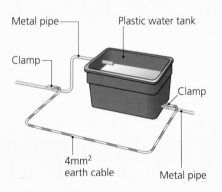

Bridge across a plastic water tank from metal pipe to metal pipe with 4mm^2 earth cable.

Link together non-electrical metal fittings in bathrooms or shower rooms with 4mm^2 earth cable. Such fittings might include inlet pipes to taps on the basin, bath and bidet, towel rails and the metal cradles of plastic baths. Connect the earth cable to the metal with clamps. Hide cable behind fittings and under the floor.

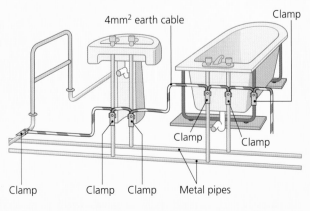

MAIN BONDING

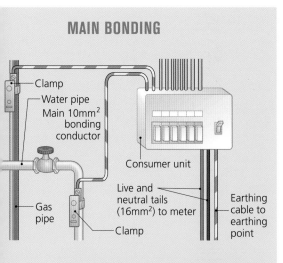

- Clamp
- Water pipe
- Main 10mm² bonding conductor
- Consumer unit
- Live and neutral tails (16mm²) to meter
- Earthing cable to earthing point
- Clamp
- Gas pipe

The main bonding conductors should be connected at one end to terminals on the earth block in the consumer unit and at the other to water, gas and oil pipes.

Clamps must fit tightly and be in contact with bare metal, not paint. They must be free from corrosion and should be easily accessible for inspection.

Additions to such circuits should follow the rules shown left and described in the next project. There is no need to run a separate earth cable from the consumer unit.

New bathroom circuits – that is those wired to the 17th Edition of the Wiring Regulations – do not require supplementary bonding provided they are RCD protected.

Installing supplementary bonding conductors

Use 4mm² green-and-yellow-insulated earth cable. To connect it to metal pipes, use clamps complying with British Standard BS951.

Plan the route of the bonding cable. You can lead the earth cable from the metal item either to the earth terminal in a socket outlet or to metal pipework which has been bonded.

If you are connecting it to a socket outlet, first turn off the main switch at the consumer unit and remove the fuse for the circuit the socket outlet is on. If you are going to connect to a pipe, make sure that all of the pipe will be bonded. The bonding continuity will be interrupted by plastic water tanks, plastic push-fit fittings and by replacement sections of plastic pipe and all of these must be bridged with 4mm² green-and-yellow insulated cable.

The connections of supplementary bonds should be accessible for inspection and checking. Connections to the earth terminals of socket outlets are easy to inspect. Connections to pipework can be made in cupboards, behind removable panels or under the floor. If they have to be made under the floor, the section of floorboard above should be fixed with screws so that it can easily be lifted if necessary.

Metal baths are fitted with an earth tag. Connect the bonding cable by winding the bared end of the conductor round a bolt passed through the tag. Trap it with a metal washer secured under the nut. Make sure that the tag is clean and free of paint or enamel.

An earth clamp Use a clamp to connect the earth cable to metal pipework. Clean the pipe with wire wool first. If the pipe has been painted, strip off an area of paintwork. Screw the core of the earth cable tightly into the terminal. The clamp should already be fitted with a permanent metal label that says SAFETY ELECTRICAL CONNECTION – DO NOT REMOVE. Make sure this is in place.

Wiring in bathrooms

Correct earthing and supplementary bonding are particularly important in bathrooms, where the combination of water, electricity and bare skin spell danger. In addition to this, the Wiring Regulations define areas (called zones) in bathrooms where specific safety rules must be followed. These describe what electrical equipment can be installed in each zone.

All new wiring work in bathrooms is notifiable to your local Building Control Department as are any modifications or additions to bathroom wiring. Because of the changes in the 17th Edition of the Wiring Regulations (see page 35), it might be better to leave all bathroom wiring to a qualified electrician.

Bathroom zones

A room that contains a bath or shower is split into three zones.

Zone 0 is the inside of the bath or shower tray. No electrical equipment is permitted here. The space beneath the bath is outside any zone if the bath panel is secured with screws and cannot be removed without the use of a screwdriver. This makes it a suitable place for the installation of equipment such as a shower pump. However, if there is no panel, or if the panel can be removed without the use of tools, the area beneath the bath is designated as zone 1.

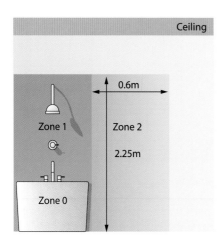

Zone 1 is the space immediately above the bath or shower tray, and extends to a height of 2.25m above the floor. Within this zone you can install: an instantaneous

shower or water heater; an all-in-one power shower unit with a waterproof integral pump, a shower pump, a towel rail, a whirlpool unit, ventilation equipment and suitable light fittings. All of these must have suitable IP ratings (see Box), especially if a power shower is fitted, which may require a minimum of IPX5 equipment.

Zone 2 extends horizontally to 600mm from the edge of zone 1 (that is, the edge of the bath or shower tray), and vertically to a height of 2.25m above floor level. Within Zone 2, you are allowed any of the equipment allowed in Zone 1 plus a shaver unit that complies with BS EN 61558-2-5.

IP RATINGS

Any equipment installed in Zones 1 or 2 must have suitable protection against water splashes. IP ratings consist of two digits. The first describes the level of physical protection the equipment provides against contact with live (or moving) parts. The second describes the level of protection against water penetration. The higher the number, the greater the protection provided. An X is used where the protection is not relevant. The minimum ratings for bathrooms are IPX7 (protection against immersion) for Zone 0 and IPX4 (protection from splashed water) for Zones 1 and 2 except for shaver units properly installed.

Outside the zones you are in theory allowed to have any equipment with the exception of socket outlets that must be at least 3m away from the boundaries of Zone 1 (which rules out sockets in all but the largest bathrooms). Most people would probably be happier to follow the previous Wiring Regulations and not have any sockets in a bathroom at all – see *17th Edition Rules*.

Bathroom switches
Wall-mounted light switches and double-pole switches (as part of a fused connection unit or for a shower unit) are not allowed within the Zones. Switches here must be the ceiling-mounted type, with an insulated pull-cord.

17TH EDITION RULES

The latest Wiring Regulations have introduced significant changes for wiring in bathrooms – some less stringent than before, some more. The main changes are:
• All new bathroom circuits must now be RCD protected. These include a shower circuit, the bathroom lighting circuit and the ring circuit that supplies power to an extractor fan, shower pump or towel rail;
• Supplementary equipotential bonding (see page 32) is no longer required (provided the circuits are RCD protected and all main services are bonded – see page 32);
• Socket outlets are now allowed, but only if they are at least 3m from the Zone 1 boundary, so cannot be touched by anyone using the bath or shower.
These changes are not retrospective, but do apply to 'extensions and modifications' to existing installations.

Tools for wiring work

Many of the jobs involved in wiring are non-electrical in nature – lifting floorboards, say. But for the electrical work, some special tools are essential.

Torch Choose one with a sturdy stand, or clip-on fitting. A powerful torch will light up work under floors and in lofts. Have a supply of spare batteries available or ideally choose a torch with a rechargeable battery.

Wire strippers The adjustable blades will strip the insulation from cores of different sizes in cable and flex without damaging the conductors inside.

Pliers A pair of 160mm combination pliers is useful for twisting cable conductor cores together prior to insertion into terminals. The cutting jaws can also be used for cutting cable and flex.

Circuit continuity tester With a simple battery-powered tester you can check the continuity of circuits and whether a socket outlet is on a ring main circuit or on a spur.

Tester screwdriver An insulated screwdriver with a 3mm blade is used for tightening terminal screws in plugs and other wiring accessories. A bulb in the handle lights up if the tip touches a live terminal or conductor.

Insulated screwdriver A larger screwdriver with an insulating sleeve on the shaft is useful for undoing and tightening plug screws and the screws fixing accessory faceplates to their mounting boxes.

Wire cutters A pair of 125mm or 160mm wire cutters will cut cable and flex, and trim conductors to length.

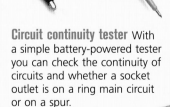

Trimming knife A sharp blade will cut through the outer sheath of cable and flex.

Changing a fuse in a consumer unit

If you still have circuit fuses, keep spare fuses or fuse wire to hand for instant repairs if a fuse 'blows'.

Mending a rewirable fuse

Tools *Insulated screwdriver.*
Materials *Fuse wire.*

1 Turn off the main on/off switch in the consumer unit. On an older system it may be in a separate enclosure near the meter. Remove or open the cover over the fuse carriers.

2 Pull out each fuse carrier in turn to find out which has blown. Scorch marks often show this, or simply a break in the wire.

3 If a power circuit is affected, switch off and unplug all the appliances on the circuit. If it is a lighting circuit, turn off all the light switches. If you do not switch everything off, the mended fuse is likely to blow again immediately you turn the main switch back on. Replace the fuse wire (right).

Replacing the fuse wire

1 Loosen the two terminal screws and remove any pieces of old wire. Cut a new piece of fuse wire of the correct amp rating, long enough to cross the carrier and go round both screws.

2 Wind the wire clockwise round one screw and tighten the screw.

3 Pass the wire across the bridge or thread it through the holder. If you are unsure about how the wire runs in the carrier, examine one of the intact fuses.

4 Wind the wire clockwise round the second screw. Make sure there is a little slack in the wire so that it will not snap and then tighten the screw. Cut off loose wire.

5 Replace the fuse carrier in the consumer unit. Close the cover and restore the power by turning on the main switch.

Checking the circuit

Look for damage on the appliances, lights and flexes that were in use on the circuit when it failed. Make repairs if necessary, then switch on the appliances or lights one at a time. Check that you are not overloading the circuit with too many high-wattage appliances. Overloading is the likeliest cause of the blown fuse. If the fuse blows again, call an electrician.

TYPES OF REWIRABLE FUSE CARRIER

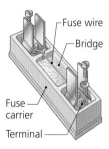

Bridged fuse
The wire runs from one terminal to the other over a plug of white arc-damping material. The carrier is ceramic.

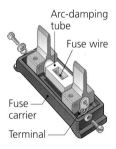

Protected fuse
Between the terminals the wire runs through a porcelain arc-damping tube. The carrier is tough plastic.

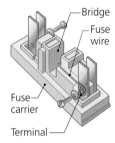

Fuse between humps
The unprotected wire passes round humps between one terminal and the other. The carrier is ceramic.

Replacing a cartridge fuse

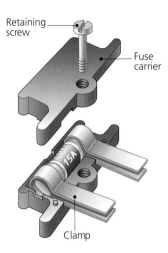

Retaining screw

Fuse carrier

15A

Clamp

Tools *Insulated screwdriver; fuse tester.*
Materials *Cartridge fuses.*

1 Turn off the main switch on the consumer unit.

2 Find out which fuse has blown unless you already know: take out each fuse carrier in turn so you can test the cartridge.

3 Prise the cartridge gently from the clamps. Some carriers are in two halves and the screw holding them together has to be removed to give access to the cartridge.

4 Test the cartridge with a fuse tester (see right). Remove only one carrier at a time. Test its cartridge and replace the carrier before removing the next one for inspection and testing.

5 When you have traced the blown fuse, replace the cartridge with a new one of the amp rating shown on the carrier.

6 As with a rewirable fuse, switch off all appliances or lights on the affected circuit. Replace the fuse carrier, close the box and turn on the main switch. Check the circuit in the same way as for rewirable fuses.

Checking a miniature circuit breaker

If the consumer unit is fitted with miniature circuit breakers (MCBs) instead of circuit fuses, it is immediately clear which circuit is affected. The switch will be in the 'off' position or the button will have popped out.

1 Turn off the main switch on the consumer unit.

2 Switch off all appliances or light switches on the affected circuit. If you do not do this, the MCB may trip off again when you reset it.

3 Push the MCB switch to the 'on' position or push in the button. Then turn the main switch back on.

4 Check the circuit in the same way as for a rewirable fuse (left).

TESTING A FUSE

You can buy an inexpensive tester that will tell you if a cartridge fuse has blown. Some types can also check flat batteries and blown light bulbs.

Choosing the right fuse

Always use the correct fuse for the job in hand. NEVER use any other metallic object or material in place of a blown fuse in order to restore power to a circuit or appliance. Doing so would remove the protection the fuse provides, and could allow an electrical fire to start or result in someone receiving a potentially fatal electric shock.

Fuse wires
If your consumer unit has rewirable fuses, use 5amp wire for a lighting circuit, 15amp wire for an immersion heater circuit, and 30amp wire for a ring main circuit or a circuit to a cooker rated at up to 12kW.

Cartridge fuses

5A Use for a lighting circuit

15A Use for a storage heater or immersion heater circuit.

20A Use for a 20amp radial power circuit and storage heater.

30A Use for a ring main circuit or a 30amp radial power circuit.

45A Use for a cooker or shower circuit.

Choosing flexes

Most flex is round in cross-section and has a white PVC outer sheath that contains colour-coded insulated conductors. The live conductor is brown, the neutral blue and the earth green-and-yellow. Each conductor (or core) is a bundle of wires. The thicker the core, the more current it can carry.

Ordinary PVC-sheathed flex will withstand temperatures of up to 60°C. Heat-resistant rubber-sheathed flex will withstand temperatures of up to 85°C. Non-kink flex has a rubber sheath with an outer cover of braided fabric. Flex with an orange sheath is used out of doors.

Metal light fittings and most appliances need three-core flex. Two-core flex with no earth core is used on double-insulated power tools and appliances (marked ▣), and for wiring non-metallic light fittings.

Key to flex colours
L – Live (brown)
E – Earth (green and yellow)
N – Neutral (blue)

2-core flex For non-metallic light fittings and double-insulated appliances. Available with flat or round PVC sheath. Use 0.5mm² for up to 700W with a pendant weight of 2kg; 0.75mm² for up to 1.4kW and 3kg; and 1mm² for up to 2.3kW and pendant weight of 5kg.

3-core flex Used for all other appliances and for metallic light fittings and pendant lampholders requiring earthing. Has a round PVC sheath. Use 0.75mm² flex for up to 1.4kW, 1mm² for up to 2.3kW, 1.25mm² for up to 3kW; 1.5mm² for up to 3.7kW. Use 1.5mm² heat-resisting flex for immersion heaters.

3-core braided flex Used for portable appliances, such as irons and toasters, with hot parts that could damage PVC-sheathed flex. Use sizes as for 3-core PVC flex.

3-core curly flex Useful for worktop appliances such as kettles, to keep flex safe and tidy. Use sizes as for 3-core flex.

Preparing flex for connection

The cores inside flex must be exposed before they can be connected to the terminals of a plug, appliance or ceiling rose.

Tools *Trimming knife; wire cutters and strippers; pliers.*

Materials *Flex. Possibly also PVC insulating tape or a rubber sleeve.*

Stripping the outer sheath

Most flex has an outer sheath of tough PVC. Remove enough to make sure that the cores can reach the terminals easily or they may be pulled out. For most connections you need to remove about 40mm of the sheath. Take care not to cut or nick the insulation on the cores as you cut the outer sheath.

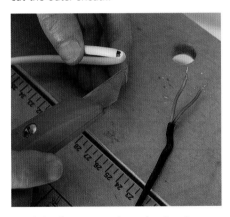

Bend the flex over and cut the sheath lightly with a trimming knife. The tension at the fold will open up a split halfway round the sheath. Fold the flex the other way and repeat. Then pull off the unwanted length of sheath.

Cutting and stripping the cores

1 Cut the individual cores to the right length to reach their terminals.

2 Set the wire strippers to match the thickness of the cores you are stripping. The core should just be able to slide out of the opening in the tool.

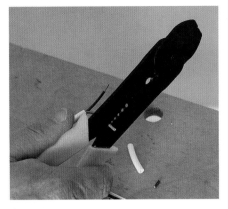

3 Press the handles together to cut the core insulation about 15mm from the tip. Rotate the strippers half a turn and pull them towards the tip of the core. The insulation will slide off.

4 Twist the strands of wire together.

Alternatively If you are preparing very thin flex for connection, strip off 30mm of insulation from each core, rather than 15mm. Twist the wire strands together, then fold the bare core over on itself in a tight U-shape. This makes it easier to insert into the terminal and provides a better electrical contact.

Fabric-covered flex

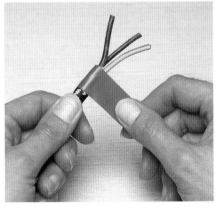

The outer cover of braided fabric on non-kink flex is likely to fray where it is cut. Wrap a strip of PVC insulating tape two or three times round the cut end of the fabric to seal down the loose threads.

Alternatively Cover the cut with a purpose-made rubber sleeve. This will be held by the cord grip of the plug. Remember to put it on before inserting the flex in the plug.

Extending a flex

Never join lengths of flex by twisting together the cores and binding the join with insulating tape. It may overheat and start a fire.

If you have to extend a flex, use a one-piece connector to make a permanent joint, or use a two-part connector if you want to be able to separate the joint. This must have three pins for connecting appliances that use three-core flex. Two-pin connectors are used mainly for connecting double-insulated garden power tools (marked with the symbol ▣) to extension leads.

Tools *Insulated screwdrivers; trimming knife; wire cutters and strippers; pliers.*

Materials *Flex connector; length of flex fitted with a plug.*

A one-piece connector

1 Unscrew the connector cover and remove it. Prepare the ends of both flexes for connection. Check that the cores (the wires) are long enough to reach the brass terminals when each flex sheath is held over its cord grip.

2 Lift out the brass barrel terminals and loosen all the terminal screws.

3 Push the cores into the terminals so that they match – brown to brown in one terminal, green-and-yellow to green-and-yellow in the second, and blue to blue in the third. Tighten all the terminal screws.

4 Loosen one screw and remove the other from each cord grip so you can swing the bar aside.

5 Fit the brass barrel terminals in their slots and position each flex sheath beneath its cord grip.

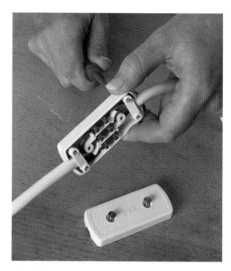

6 Replace the cord grip screws and tighten them to grip the flex sheaths securely. Fit and screw on the cover.

A two-part connector

You must fit the flex from the appliance to the part of the connector with the pins, and the flex from the mains supply to the part with the sockets.

If you fit the flexes the other way round, you will have live pins exposed if the two parts of the connector become separated while the power is switched on.

1 Prepare the ends of both flexes. Separate the two halves of the connector. Remove the screw holding the terminal block inside each cover and push it out. Undo the cord grips.

2 Slide the outer covers onto the two prepared flexes.

3 Connect the cores to each terminal block with the green-and-yellow core in the middle. Make sure that the brown and blue cores in each part of the connector are opposite each other so they will connect when the parts are joined.

Extension leads and adaptors

Trailing socket adaptor
This provides three or more outlets
and is normally fitted with a fuse and
neon indicator. It plugs into an existing socket
and can supply a total wattage of up to 3kW. Use
for low wattage appliances like computer equipment.

Plug-in adaptor Allows two
or three plugs to fit into a
single socket outlet. The total
wattage for the adaptor is
3kW and it should be fitted
with a 13A fuse. Do not plug
one adaptor into another;
the pin contacts will be poor
and overheating could result.

Wire-in adaptor Makes a permanent connection for up to
four low-wattage appliances and supplies total wattage of
up to 3kW. Mount on the wall near the socket. Use for hi-fi
or other equipment that is always kept in the same place.

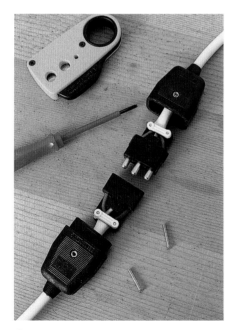

4 Fit each flex sheath in its cord grip and
tighten the screws to hold it securely. Push
each terminal block back into its cover and
replace the fixing screw.

Choosing an extension reel

**Even if your home has a sufficient
number of socket outlets, there will be
times when you need to use an
extension reel. This allows you to have
power down the garden, up on the
roof, up in the loft; anywhere where
there is not a convenient socket outlet.**

Many power tools have flexes so short that
you will need an extension lead almost
every time you use them. A cable reel has
the advantage that you get two, three or
four sockets on the reel – but note that the
total load for all the sockets is equal to the
load current rating of the reel.

An extension reel has two ratings – one
for the cable fully wound up and the other
for it fully unwound. For most DIY, it's best
to have a 13amp (13A) reel (the highest
rating) and always to use it fully unwound
– it's possible to 'cook' a lower-rated reel if
you use it unwound. There are three main
types of reel: cassette reels, 'handbag' reels
and heavy-duty stand-alone open reels.

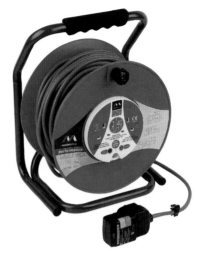

Cassette reels These are the cheapest type and come in lengths up to 15m, with typical ratings of 6A, 10A and 13A. Better suited to use inside the house: look for reels with a thermal cut-out (see *Reel Safety* below), a built-in hand grip and a rotating knob for easier winding.

Open reels An open plastic drum on a galvanised steel frame with a rubber carrying handle. Lengths from 20m to 50m, mostly fitted with thermal cut-outs and some with RCD plug (see *Reel Safety*, below). On some stand-alone reels, the sockets do not turn with the reel, so the reel can be extended with the plugs in place. Look for locking brakes, cable tidies and ON indicator lights.

REEL SAFETY

Thermal cut-out Many extension reels are fitted with a thermal cut-out that operates if the flex on the reel gets too hot. You have to wait until the reel has cooled down before re-setting the cut-out.

RCD protection It is essential that an extension reel should always be used with an RCD, which cuts off the power at once if you sever the flex and touch a live core.

Unless the socket outlet into which you plug the extension lead is on a circuit that is RCD protected (see page 17) or the socket outlet itself has an RCD built in (see page 120 for an example), you should always use an RCD adaptor (left) between the plug and the socket outlet. Some extension reels are fitted with an RCD plug that can safely be plugged into any socket outlet.

'Handbag' reels A cassette reel in a plastic case (with a handle) that can be stood up, making the reel suitable for use outside. Mostly 13A, with lengths from 15m to 50m. Some reels fitted with RCD plug (see *Reel Safety*, left) as well as thermal cut-out. Again look for rotating knobs; some handbag reels have a place for storing the plug and some have ON indicator lights.

Choosing plugs and fuses

Plastic plug Moulded plastic plugs are the commonest type. They are usually white, but other colours are available. All new plugs have plastic sleeves on the live and neutral pins, to prevent accidental finger-tip contact with live metal as you pull the plug out of its socket outlet.

Moulded-on plug
All new appliances should have a one-piece factory-fitted plug. These cannot be opened, so must be cut off and replaced if damaged. Hammer the pins out of line before you throw it away so it cannot be plugged into a socket outlet if found by a child. The cut end of the flex would give a shock if touched.

Rubber plug
Tough rubber plugs are intended for use on power tools. Rubber will not crack if the plug is knocked or dropped in use. You have to thread the flex through the rubber cover before connecting it to the plug terminals.

3amp cartridge fuse
Use in plugs and fused connection units for appliances rated at up to 700 watts. Check the wattage on the rating plate of the appliance. Low wattage appliances include table or standard lamps, hi-fi equipment, home computers and ancillaries, and electric blankets.

13amp cartridge fuse
Use in plugs and fused connection units for appliances between 700 and 3000 watts such as most TV sets, vacuum cleaners, large power tools, room heaters and all domestic appliances that contain a heating element.

Fitting a new plug

All electrical appliances sold in the UK must have a factory-fitted plug. This has greatly improved household electrical safety, by eliminating the need for the consumer to fit a plug to every new appliance – a task that many found difficult to carry out correctly.

However, you will need to fit a replacement plug if the factory-fitted one is damaged. Many older appliances in the home will still have hand-wired plugs, which may also need replacing over time.

The colours of the plastic insulation on the cores in flex were changed to brown (live), blue (neutral) and green-and-yellow (earth) in 1968. Any appliance with flex cores coloured red (live) and black (neutral) should be checked for electrical safety.

All three-pin plugs are fitted with a cartridge fuse. Many contain a 13amp fuse when you buy them, but you should fit a lower-rated fuse if the appliance rating is below 700 watts (see Choosing plugs and fuses, page 43).

Old colours	New colours
E ⏚ N L	E ⏚ N L
Black to N Green to earth (E or ⏚) Red to L	Blue to N Green-and-yellow to earth (E or ⏚) Brown to L

Tools *Insulated screwdrivers; trimming knife; wire cutters and strippers; pliers.*

Materials *Plug; cartridge fuse (either 3amp or 13amp).*

1 Unscrew the cover of the new plug and remove it.

2 Prise out the cartridge fuse if necessary to reveal the terminal. Loosen the screw-down bar that secures the flex if there is one. Nylon jaws grip the flex in some plugs.

3 If you are replacing a hand-wired plug, remove its cover and loosen the terminal screws to release the flex cores from their terminals. Release the flex from the cord grip. Inspect the bare cores. If they appear damaged, cut them off and strip off some core insulation to expose undamaged wires ready for reconnection to the new plug.

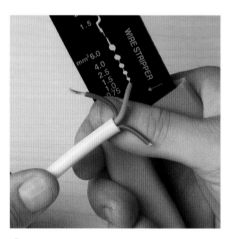

4 If you are replacing a factory-fitted plug, cut through the flex close to the plug body. Prepare the end of the cut flex (page 39). For some plugs all the cores have to be the same length, for others they have to be different lengths. Check that the prepared cores are long enough to reach their terminals with the flex sheath held in the flex grip.

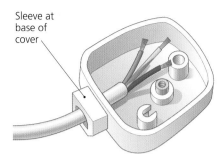

Sleeve at base of cover

5 Tough rubber plugs designed for use on power tools have a hole in the plug cover through which the flex passes before being connected to the plug terminals.

6 Connect each flex core to its correct terminal. The **BR**own (live) core goes to the **B**ottom **R**ight terminal, the **BL**ue (neutral) core to the **B**ottom **L**eft terminal, and the earth core in three-core flex (green-and-yellow) to the top terminal.

7 With pillar-type terminals, loosen the terminal screw and insert the bare end of the core in the hole. Tighten the screw to trap it in place. Plugs with this type of terminal often have loose pins; remove these from the plug first if it makes connecting the cores easier.

Alternatively With screw-down stud terminals, remove the stud and wind the bare end of the core clockwise round the threaded peg. Screw the stud down to trap the wires in place.

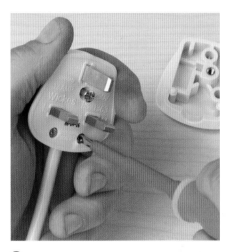

8 Arrange the cores in their channels in the plug body and place the flex sheath in the cord grip. If the plug has nylon jaws, press the flex in between them. If it has a screw-down bar, undo one screw, position the flex in the grip, swing the bar back over it and screw it down securely. Fit the fuse.

9 Replace the plug cover and make sure that it is firmly screwed together.

Reconnecting flex to a lampholder

The flex connections within a pendant lampholder may pull away from their terminals in time and stop the light working, but it is a simple matter to reconnect them.

• You will also need to reconnect the cores if you are shortening the flex, perhaps to fit a new lampshade that needs to be higher.
• If the lampholder is a metal one without an earth terminal, replace it with an earthed one – or with a plastic lampholder if the flex has no earth core. You must use three-core flex with a metal lampholder or metal lampshade.
• If the flex is discoloured or cracked, replace it with new flex.

The parts of a lampholder

Flex

Upper cover

Neutral core

Terminal

Live core

Body of holder

Slot for bayonet cap bulb

Retaining ring to hold up lampshade

Points to contact plungers inside body of holder

Light bulb with bayonet cap

Tools *Insulated screwdrivers, one with a small, fine tip; trimming knife; wire cutters and strippers; pliers.*

Materials *Replacement flex. Perhaps a new lampholder.*

1 Turn off the power at the consumer unit and remove the fuse or switch off the MCB protecting the circuit that you will be working on.

2 Remove the light bulb and unscrew the ring that holds the lampshade. With a very old lampholder, you may have to break the ring by crushing it with pliers; some shops sell the new rings separately. Remove the shade.

3 Unscrew the upper cover of the lampholder and push it up the flex to reveal the flex connections. Unhook the flex cores from the lugs on the body of the lampholder.

4 With the fine-tipped screwdriver, undo the terminal screws enough for you to draw out the flex cores.

5 The light bulb makes its connection via the two spring-loaded plungers in the base of the lampholder. Push these in to see whether they return to their original positions when released. If they do not, or if the lampholder is damaged or scorched, fit a new one.

6 Prepare the ends of the flex cores for connection (page 39).

7 If you are fitting a new lampholder, thread the new cover onto the flex.

8 Screw the brown and blue cores tightly into the lampholder terminals. It does not matter which each goes to. In a metal lampholder you must connect the green-and-yellow earth conductor to the earth terminal.

9 Hook the flex cores over the support lugs. Screw the lampholder cover on, taking care not to cross-thread it.

10 Insert the lampholder into the lampshade and screw the retaining ring in place to secure the shade. Fit the light bulb.

11 Replace the circuit fuse or switch on the MCB and restore the power at the consumer unit.

Connecting flex to a ceiling rose

If a light flex has discoloured or become brittle, it is easy to connect a new one between the ceiling rose and the lampholder.

Before you start Inside a modern ceiling rose on a loop-in wiring system (page 102) is a row of terminals in three groups. The live and neutral conductors of the circuit and switch cables are connected to these. A separate terminal is marked E or ⏚ for the earth conductors.

Use the right type and size of flex for the installation (page 38). If it connects with a metal lampholder or light fitting it must have an earth core.

Tools *Insulated screwdrivers, one with a small, fine tip; trimming knife; wire cutters and strippers; pliers.*

Materials *Replacement flex.*

Disconnecting the old flex

1 At the consumer unit, turn off the power and remove the fuse or switch off the MCB protecting the circuit you will be working on. It is not enough simply to turn off the light switch; the cables in the ceiling rose will still be live.

2 Remove the light bulb and shade to avoid the risk of dropping them.

3 Unscrew the cover of the ceiling rose and slide it down the flex.

4 Using the small screwdriver, loosen the terminal screws connecting the flex cores at each end of the row of terminals. Withdraw the cores from the terminals and unhook them from the lugs.

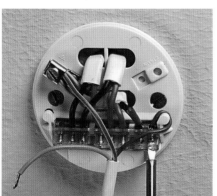

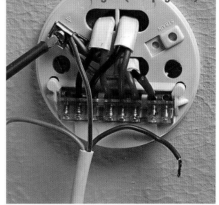

5 If the flex has an earth core, unscrew the earth terminal enough to withdraw it. Do not dislodge the other cable earths.

Connecting the new flex

1 Connect the new flex to the lampholder as described opposite.

2 Thread the new flex through the cover of the ceiling rose.

3 Prepare the new flex for connection (page 39). Take care not to strip off too much of the outer sheath. The cores have to reach the terminals without strain, but they must not show below the ceiling-rose cover; the outer sheath of the flex must enter the hole in the cover.

4 Slip the end of the green-and-yellow-insulated earth core into the earth terminal in the ceiling rose. Make sure before you tighten the screw that the other cable earth cores from the circuit and switch cables have not been dislodged from under the terminal screw.

5 Connect the blue flex core to the terminal where the circuit cable neutral (black) cores are connected.

JUNCTION BOX SYSTEM

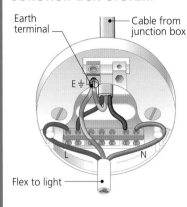

Earth terminal — Cable from junction box

E ⏚

L   N

Flex to light

A loop-in ceiling rose might have been used on a junction box system (page 102). There will be only one cable entering the rose base. The flex connection is the same as for a loop-in system (steps 5–7, pages 47–48).

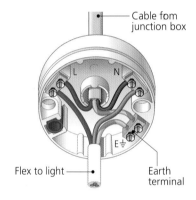

Cable fom junction box

L   N

Flex to light   Earth terminal

E ⏚

Another type of ceiling rose, used on a junction box system (above), has three sets of separate terminals, not in line.

1 Connect the blue flex core to the same set of terminals as the black cable core.

2 Connect the brown core from the flex to the same set of terminals as the red core from the cable.

3 Connect the green-and-yellow earth core from the flex to the same set of terminals as the earth core from the cable.
   If the earth from the cable has not already been sleeved with green-and-yellow insulation, disconnect it and sleeve it before connecting it – with the flex earth – to the terminal.

6 Connect the brown flex core to the terminal at the other end of the row where the switch cable neutral core is connected. This core is usually identified with a strip of red PVC insulating tape, to show that it is, in fact, live.

7 Hook the flex cores over the support lugs at each side of the rose baseplate.

8 Slide the rose cover up and screw it onto the baseplate. Replace the shade and bulb.

9 Replace the circuit fuse or switch on the MCB and restore the power.

Rewiring a table lamp or standard lamp

To fit a new flex to a table or standard lamp with no exposed metal parts, or for a double insulated lamp (marked ▣), use two-core flex without an earth core. For a lamp with metal parts, use three-core flex.

Tools *Insulated screwdrivers; trimming knife; wire cutters and strippers; pliers.*

Materials *Lamp; suitable flex. Perhaps also an in-line flex switch.*

1 Unplug the lamp and remove the bulb.

2 With a plastic lampholder, unscrew the upper cover to release the shade. With a brass lampholder, unscrew the narrow ring.

3 With a plastic lampholder, unscrew the body from the base, then screw down the lower cover to reveal the terminals.
   With a brass lampholder, unscrew and remove the second wide ring so that you can lift off the outer lampholder section and then pull out the inner section which has the flex connected beneath it.

4 Release the flex cores from the terminals. Wind the cores securely round the end of the new flex and tape the two together.

5 Gently pull out the old flex from below, using it to pull the new flex through the lamp base.

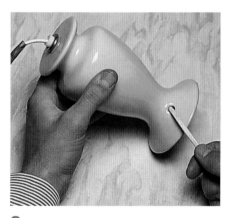

6 Prepare the flex for connection (page 39) and then finish drawing it through the lamp base until only about 40mm is protruding.

7 Connect the brown flex core to one lampholder terminal and the blue flex core to the other. For a brass lampholder, connect the green-and-yellow earth core to the E (or ⏚) terminal on the lower cover.

8 With a plastic lampholder, screw the lower cover over the terminals and screw the lampholder body onto the lamp base. As you do so, turn the flex or it will become twisted. With a brass lampholder, lower the inner section of the lampholder into place. Fit the outer section on top and secure it with the wide screw-on ring.

9 Replace the lampshade and secure it with the upper narrow ring or plastic upper cover. Fit the light bulb.

HANDY TIPS
• If the lamp has a brass lampholder without an earth terminal, you must fit a new brass lampholder with an earth terminal, or a plastic lampholder. If this is not possible, do not use the lamp.
• The flex must be threaded up inside the lamp base. The hole to thread it through is often on the side of the lamp near the bottom.
• If the hole is underneath, the lamp base should have small feet to raise it and keep its weight off the flex.
• Most lamps have a push-through switch in the lampholder. If there is no integral switch, fit an in-line switch in the flex (see page 50).

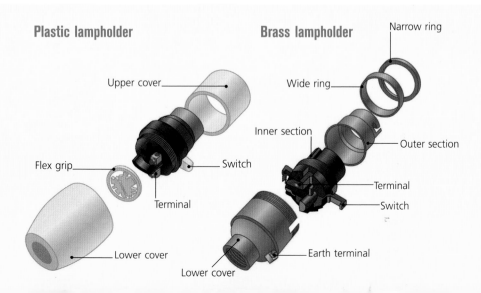

Plastic lampholder — Upper cover, Flex grip, Switch, Terminal, Lower cover

Brass lampholder — Narrow ring, Wide ring, Inner section, Outer section, Terminal, Switch, Earth terminal, Lower cover

Fitting an in-line flex switch

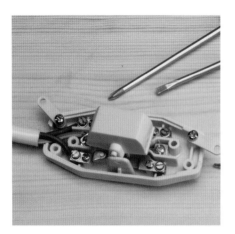

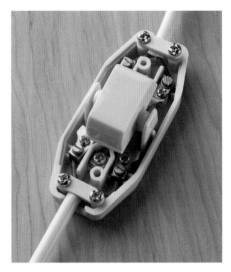

• If you are using three-core flex, use an in-line switch with an earth terminal.
• For a two-core flex on a double-insulated lamp, no earth terminal is necessary.

1 Switch off and unplug the lamp. Cut the flex where the switch is to go and prepare the ends for connection (page 39).

2 Unscrew and remove the cover of the in-line switch.

3 Take out a screw from each flex clamp so that you can swivel the clamps aside.

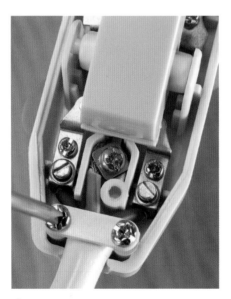

4 Release the terminal screws and connect each flex core – brown (live) to the terminals marked L, blue (neutral) to the terminals marked N, and green-and-yellow to the terminals marked E (or ⏚). Replace the flex clamps and tighten the screws.

Alternatively For a two-core flex on a double-insulated lamp, connect the two brown cores to one pair of terminals and the two blue cores to the other pair. Ignore the earth terminal.

5 Screw the switch cover back into place.

Fitting a plug to the new flex

Connect the plug as described in *Fitting a new plug*, page 44.

Replacing sockets

Electric shock or fire could be caused by contact with the conductors in a damaged socket. Replace it as soon as possible.

Replacing a faceplate

If a socket outlet faceplate is burnt or cracked by an impact, do not use the socket until you have replaced the faceplate. If the socket outlet is surface-mounted, replace the mounting box as well if it is damaged. You could even replace a single socket with a double one at the same time.

Tools *Large and small insulated screwdriver; perhaps also a trimming knife; wire cutters and strippers; pliers.*

Materials *New socket outlet. Perhaps a new plastic mounting box, with fixing screws; green-and-yellow earth sleeving.*

1 Turn off the power at the consumer unit and switch off the MCB that protects the circuit you want to work on (or remove the appropriate fuse if you have an old-style fuse box). Plug in and switch on a lamp you know to be working to check that the socket outlet is dead.

2 Undo the screws that hold the faceplate in place and pull it away from the wall. Keep the screws; you may find that the ones supplied with your new socket outlet will not fit the lugs in the existing mounting box. In that case, use the old screws.

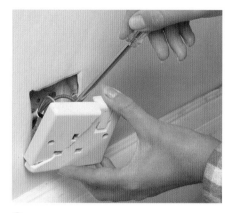

3 Loosen the three terminal screws on the back of the faceplate to release the cores. Note how many cores are connected to each terminal; there may be one, two or three depending on whether the outlet is on a spur, is on the ring circuit or is supplying power to an extra outlet on a spur. The new outlet must be connected in the same way as the damaged one.

4 If there is damage to the plastic mounting box of a surface-mounted socket, undo the screws holding it to the wall and remove it. Pass the cable or cables through one of the knockouts of a new surface-mounting box and screw the box in place.

5 Check that the cable cores will reach the terminals on the new faceplate. If necessary, cut back the cable sheath with a trimming knife to expose longer cores, taking care not to nick the core insulation. If the earth cores are bare, cover them with green-and-yellow plastic sleeving, leaving only the metal ends exposed.

6 Insert the cores into their terminals and tighten the terminal screws firmly. Connect the red core(s) to the terminal marked L, the black core(s) to the terminal marked N, and the earth cores to the terminal marked E or ⏚. Tug each core to check that it is securely held by its terminal screw.

7 Place the faceplate over the mounting box, folding the cables carefully into it.

8 Screw the faceplate to the box. Do not over-tighten the screws or the plastic may crack.

9 Turn on the MCB for the circuit you have been working on, or replace the fuse you removed. Restore the power and check that the new socket outlet works – by plugging in a lamp, for example. If it does not, switch off again at the mains and check and tighten the connections.

Replacing a single socket outlet with a double

If the single socket outlet is surface-mounted, use a double surface-mounted replacement. If it is flush-mounted (or recessed), you can fit a plastic surface-mounted box called a pattress over it. Alternatively, you can remove the existing metal mounting box, enlarge the recess and fit a larger box. Flush-mounted fittings are neater, and safer because they are less likely to be damaged by knocks from furniture. You will have to cut into the wall to fit a new double or triple flush mounting box.

Tools *Suitable tools for enlarging the recess (page 57); insulated screwdriver; trimming knife; wire cutters and strippers; pliers.*

Materials *Double or triple mounting box (or pattress, see opposite); new socket outlet; rubber grommet; screws and wall plugs. Perhaps earth sleeving.*

● Cable core colours have changed (see page 56)

MAINTENANCE AND REPAIRS

Removing the single outlet

1 Turn off the main switch on the consumer unit and switch off the MCB that protects the circuit you will be working on. Remove the appropriate fuse if you have an old-style fuse box. Check that the socket outlet is dead by plugging in a lamp you know to be working.

2 Unscrew the faceplate and ease it out. Disconnect the cable cores from their terminals and set the old faceplate aside.

Fitting the mounting box

Flush-mounted boxes

1 Take out the screws holding the old box in place and ease the box out of its recess.

2 Enlarge the hole for the new double mounting box (page 58).

3 Remove a knock-out (a pre-cut entry disc) from the new box. Fit a grommet in the hole to prevent the cable sheath from chafing on its edges, then feed the cable into the box and fix the box in place.

Surface-mounted boxes/pattresses

1 If you are replacing a surface-mounted box, take out its fixing screws and remove it. If you are fitting a pattress, leave the old flush metal mounting box in place.

2 Hold the mounting box in place, check with a spirit level that it is horizontal, and use a bradawl to pierce the wall through the fixing holes.

3 Drill holes at the marked spots and insert wallplugs in them.

4 Feed the cable through a knockout in the back of the box or the hole in the pattress. Screw the box to the wall or the pattress to the old flush metal mounting box.

Connecting the double outlet

1 Use pliers to straighten the ends of the cable cores. Cover the earth cores with green-and-yellow sleeving if they are bare.

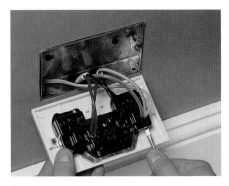

2 Screw the live (red) cores into the terminal marked L. Screw the neutral (black) cores into the terminal marked N. Screw the earth conductors into the terminal marked E (or ⊕).

3 If the mounting box is metal, fit a short length of earth core (taken from a cable offcut and covered with green-and-yellow sleeving) between the faceplate earth terminal and the earth terminal in the box. This so-called flying earth ensures that the metal box and the faceplate securing screws are safely connected to earth.

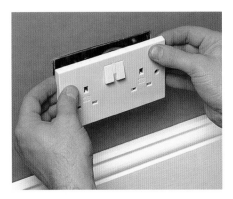

4 Fold the cables neatly into the mounting box and push the faceplate back into position over the box.

5 Screw the faceplate to the box or pattress. Do not overtighten the screws if fixing to a plastic box, or it may crack.

6 At the consumer unit, replace the fuse or switch on the MCB to restore power to the circuit you have been working on.

Choosing socket outlets and mounting boxes

There is a wide variety of socket outlets and mounting boxes to choose from. When increasing the number of outlets you have, or replacing old outlets, try to ensure that what you choose will supply enough power and sockets for any future use of electric appliances.

Plastic socket outlets

Socket outlets are available in single, double and triple configurations, with or without switches. Holes for the plug pins have spring-loaded shutters to prevent children from poking anything into them. Most sockets fit into a 32mm-deep white plastic surface box or a 25mm steel flush-mounted box; deeper (35mm) flush boxes allow more space for the wiring.

Metal socket outlets

Single and double brass, chrome and stainless-steel sockets are available in a variety of styles. There are plastic inserts round the shuttered pin holes and switches. Metal sockets fit on a flush steel box or a matching metal surface box.

Socket outlets with indicators

Both plastic and metal sockets are available with a neon indicator above each outlet. The indicator lights up to show when the socket outlet is switched on. Take care not to dislodge the small neon bulb clipped behind the coloured window when handling the faceplate.

Fused connection units

Fixed appliances, such as wall-mounted heaters and fans, should be connected to a ring circuit by a fused connection unit (FCU) with a direct flex connection, rather than by a plug. The flex from the appliance can enter the FCU through the edge or a hole in the front. The FCU may have a switch and a neon indicator, and is fitted with a fuse to match the appliance wattage.

Flex-outlet plates

In bathrooms where socket outlets are not permitted, a fixed appliance such as a towel rail is wired into a flex-outlet plate. The cable (from a fused connection unit) goes into the back of the mounting box and the flex enters through the front of the plate and is connected to terminals on the back of it. There is no switch.

Boxes for flush mounting

Metal boxes for single, double and triple socket outlets have pre-cut discs to allow cables to enter. Knock out a disc and fit a grommet in the hole to protect the cable. The earth terminal in the box must be linked to the earth terminal on the accessory faceplate with sleeved earth core. Single and double plastic 'drylining' boxes are available for use in plasterboard walls. They also have knock-out cable entry points, but do not need grommets.

Boxes for surface mounting

Single, double and triple surface boxes are made in plastic. They have knock-out cable entry points, but do not need grommets. Metal surface boxes are also available; they are used mainly in outbuildings.

Pattress for converting a socket outlet

A mounting frame called a pattress allows a surface-mounted double or triple socket outlet to be fitted over a single flush-mounted metal box.

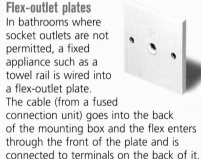

Working on electrical circuits

Many electrical jobs involve working on the house's electrical circuits. To ensure safety and success you must plan the job carefully.

A typical project will involve running cable from one point to another, either from the main consumer unit or from a point on an existing circuit. To minimise disruption, do as much work as possible with the power supply on. You can bury new cables in walls, run them beneath floorboards or through plastic trunking and create recesses for new mounting boxes for sockets with the power live. Leave the new cables sticking out where they terminate at a new socket outlet or light fitting and where they link to the existing circuit and turn off the power supply only when you are ready to make the final connections.

Extending circuits

Jobs such as adding extra socket outlets, installing a cooker hood or providing new lighting points all involve adding branch lines (called spurs) to existing circuits. A spur from a ring main circuit can supply one single, double or triple socket outlet or a single fused connection unit providing a sub-circuit to a cooker hood, extractor fan, waste disposal unit, wall heater or central heating control. Before you add a spur to a lighting circuit, check that this will not overload the circuit, which can supply lights with a maximum total of 1200 watts.

1 You can extend a ring main circuit in two ways. The spur cable can be connected into the terminals of any socket outlet or fused connection unit (FCU) that is on the main circuit. This is often the most practicable solution. Note that only one spur can be connected to each outlet or FCU.

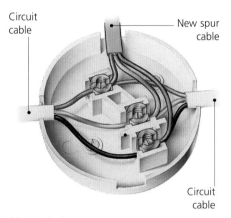

Circuit cable

New spur cable

Circuit cable

Alternatively Make a connection via a three-terminal junction box inserted in the main circuit cable. This means locating cable below floorboards but may offer a more convenient connection point if boards have to be lifted anyway to route the spur cable to its destination.

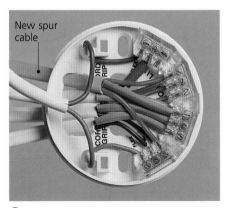

New spur cable

2 You can extend a lighting circuit in three ways. All require access to ceiling voids or the loft space. The spur cable can be connected directly into the terminals of an existing loop-in ceiling rose.

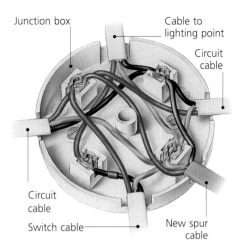

Junction box

Cable to lighting point

Circuit cable

Circuit cable

Switch cable

New spur cable

Or you can connect the spur cable into a four-terminal junction box that is already supplying an existing lighting point.

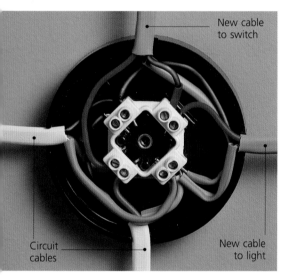

New cable to switch

Circuit cables

New cable to light

Alternatively You can make the connection via a four-terminal junction box inserted in the main lighting circuit cable. This box will supply the new light fitting and its switch. Care must be taken with this method to identify the circuit cable correctly.

Adding circuits

Jobs such as installing an electric shower or cooker require an entirely new circuit, run from the consumer unit. The circuit cable runs first to a double-pole (DP) isolating switch, then on to the appliance. For an electric shower, the double-pole isolating switch is usually ceiling-mounted and must be cord-operated for safety. The circuit then runs from here on to the shower unit, where it is connected to the appliance's terminal block.

In most situations there will not be a spare fuseway available for the new circuit cable, so one will have to be provided. There are two options for doing this, bearing in mind that the new circuit will almost certainly have to be RCD protected – all new bathroom circuits (page 34 and 66), all new socket outlet circuits (page 76) and any other circuits where the wiring in walls is not mechanically protected (pages 57-63).

1 Fit a four-way enclosure with a 63amp/30mA RCD and two MCBs (the current rating of the RCD should equal the sum of the MCB ratings) so that you can add another circuit later on. In either case, you can install the new unit alongside the existing fusebox or consumer unit and connect the new circuit cable to it (see page 122). Call in your electricity supply company or a qualified electrician to connect the new unit to the main incoming power supply; you are not allowed to do this yourself.

2 Replace the existing fusebox or consumer unit with a new, larger consumer unit to which existing and new circuit cables can be connected. This has several advantages. The new unit will provide a single main on-off switch, RCD protection for any circuit and can include some spare MCBs for future extensions to the system.

A '17th Edition' consumer unit has two RCDs with the circuits split between them, so that you are left with some light in each room when an RCD trips.

Replacing a consumer unit is a job for a qualified electrician, who will specify its contents correctly for your requirements and reconnect the unit to the incoming supply.

**New cable colour key
(effective April 2006)**
Live (brown), Earth (bare wire), Neutral (blue)

Old cable colour key
Live (red), Earth (bare wire),
Neutral (black)

1.0mm² Two-core-and-earth, used for lighting circuits.

1.0mm² Three-core-and-earth, used for wiring between two-way switches.
Cores are colour-coded for identification.

2.5mm² Two-core-and-earth, used for ring main circuits, 20amp radial socket outlet circuits, circuits for immersion heater, or storage heaters up to 20amp.

4mm² Two-core-and-earth, for 30 (or 32 amp) socket outlet circuits.

6mm² Two-core-and-earth, used for 30amp circuit for cooker up to 12kW, or 45amp circuit for shower up to 8kW.

10mm² Two-core-and-earth, used for circuit for cooker above 12kW or shower rated above 8kW.

Sleeving for earth conductors When you strip off the outer sheath of a cable and prepare its conductors for connection, you must always cover the bare earth conductor with a length of green-and-yellow PVC sleeving of the appropriate size.

Choosing cable for indoor circuits

Cable is used to wire up the circuits to ceiling roses, light switches, socket outlets and fixed appliances. It has a flattened oval cross-section, with a white or grey PVC outer sheath to protect the conductors inside.

• The most commonly used type has two conductors insulated in coloured PVC and a bare earth conductor, and is known as two-core-and-earth or twin-and-earth cable. Cable with three insulated conductors plus a bare earth is used solely for wiring between two-way switches (see page 85).
• Cable is made in several sizes, each identified by the cross-sectional area of the conductors in square millimetres (mm²). Each circuit is wired in cable sized to match its current demand.

The conductors carrying the current in cable are colour-coded so they can be connected up correctly. In existing two-core-and-earth cable, the red core is used as the live conductor, and the core coloured black as the neutral conductor. Since April 2006, cable with old core colours has not been available. All new wiring work in the home must now be done using cable with new core colours (see left). In two-core-and-earth cable, the live core is brown and the neutral core blue – the same as the colours used on flex cores, which were changed in 1968. The cores in three-core-and-earth cable are brown, grey and black. New PVC cable has a grey sheath.

There is no requirement to change existing wiring. Take care when making connections to existing wiring to link new brown cores to old red ones, and new blue cores to existing black ones. A warning notice must be fixed close to the consumer unit with the following wording:
CAUTION This installation has wiring colours to two versions of BS7671. Great care should be taken before undertaking extension, alteration or repair that all conductors are correctly identified.

Preparing the circuit route

New cable can be run under floorboards and up and down walls – either buried in a 'chase' cut into the plaster, or hidden behind plasterboard or in surface-mounted trunking.

• You will have to lift floorboards to run cables in floor or ceiling voids. If the floor is solid, you will have to run it in trunking. Do not attempt to cut chases in a concrete floor; you are likely to damage the damp-proof membrane in the structure.
• Cable buried in plaster must be protected against penetration by screws or nails – at least by oval PVC conduit, but preferably by metal conduit or galvanised capping. See *17th Edition Rules* for more details.

Tools *Electronic detector (see page 11); power drill; masonry drill bits; flat wood bits; twist drill bits; club hammer; brick bolster; cold chisel; claw hammer; pencil; trimming knife; hacksaw; chisel; spirit level; bradawl; filling knife; wire cutters and strippers; pliers; screwdrivers.*

Materials *Cable; cable clips; oval PVC conduit or metal capping; galvanised nails for fixing conduit; mounting boxes with screws; wallplugs and grommets; ceiling roses; plaster or interior filler; timber offcuts; nails.*

17th EDITION RULES

The latest Wiring Regulations have new rules for new cables that you bury or conceal in walls at a depth of less than 50mm. Even if they are run – as they should be – in the 'safe zones' (within 150mm of the top of the wall or the junction between walls or vertically/horizontally to an accessory), cables must either be robustly physically protected or the circuit of which they are part be protected by a 30mA RCD.

This means that if you are adding a new circuit or extending a non RCD-protected circuit, you may also need to add RCD protection. See pages 55 and 72.

Fitting a flush mounting box

1 Hold the mounting box in position. For a socket outlet it should be at least 450mm above the floor or 150mm above worktop level. For a light switch, it should be at about shoulder height. Check with a spirit level that the box is horizontal and draw a line round it as a guide for drilling.

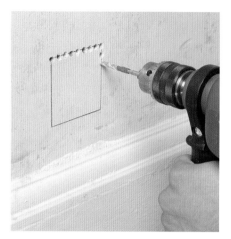

2 Using a masonry bit, drill holes all round the marked outline to the depth of the box. If the drill has no depth gauge, mark the bit with adhesive tape at the required depth and drill until the tape reaches the wall. Drill more holes within the marked area.

3 Cut out the recess to the required depth with a brick bolster and club hammer. Brush out the recess, fit the mounting box and make marks with a pencil or bradawl through the fixing holes at the back of the box. Take out the box, drill holes at the marks and insert wallplugs.

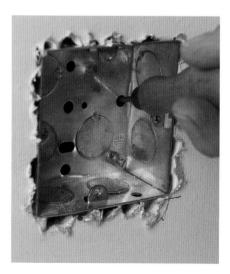

4 Knock out one of the discs stamped in the back or sides of the box for the cable to enter. Fit a grommet in the hole. Do not fix the box until the cable has been run.

Laying cable under a floor

The cable can run alongside one joist or it can cross several.

1 Where the cable is to be at right angles to the joists, lift one or two boards to get access to the joists. Drill holes through the joists at least 50mm down from the top edge and big enough for the cable to pass through easily. If necessary, drill the holes at a shallow angle.

2 Thread the cable through the drilled holes. Leave a little slack between the joists.

3 Where the cable runs alongside a joist, lift a floorboard (or a section of one) about every 500mm along the joist. Feed the cable beneath the boards and secure it with clips to the side of the joist. Run it 50mm below the top edge of the joist and hammer in a clip where the cable is exposed.

Burying a cable in plaster

1 Plan the route for the cable. It should run vertically above or below a socket outlet or switch. A horizontal run should be close to the ceiling or between sockets. Never run it diagonally.

2 Mark the route with two lines 25mm apart. Avoid making sharp bends, wherever possible.

3 Check with a pipe/cable detector (page 11) along the route to make sure that you are not going to interfere with any cables or pipes buried in the wall. If you are at all uncertain whether there is live wiring near the spot where you are working, switch off at the mains until you have made sure.

4 When you know that the route is safe, use a trimming knife to score along both edges of the chase.

5 Use a cold chisel and club hammer to cut out the plaster. Protect your eyes with safety goggles.

Alternatively Hire an electric chasing machine to cut out the chase. This creates a great deal of dust which spreads everywhere, so shut all doors.

6 Use a long masonry drill bit to make a hole behind the skirting board where cable is being taken down to run under the floor. Enlarge it with a cold chisel if necessary.

7 Cut a 3m length of oval PVC conduit to size and check that the chase is deep enough for it to fit easily. If it is too near the surface, only a skim of plaster or filler will cover it when you make good and it will probably crack. A covering of 5mm of plaster or filler over the conduit should be thick enough to prevent cracking.

Alternatively Use galvanised steel capping to protect cable. This is sold in 2m lengths, cut with a hacksaw and secured with galvanised nails. It needs a wider chase.

8 Feed the cable into the conduit and secure it in the chase with galvanised nails on each side. When the cable comes up behind a skirting, feed it into the conduit so that the end of the conduit is below the top of the skirting.

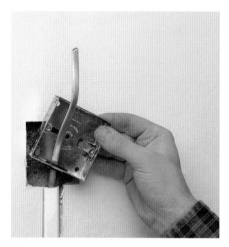

9 Leave enough spare cable at each end of the run to reach the mounting boxes easily. Ease the end of the cable or cables through the grommet into the mounting box. Slide the box into the recess and screw it into the wall behind.

Routes in stud partition walls

Cutting the route
With walls of plasterboard fixed to both sides of a timber frame, the cable can be run in the cavity behind the plasterboard. You will have to cut notches in the frame.

1 Use a bradawl or a stud detector (page 11) to locate the frame, or a pipe/cable detector to locate the rows of nails holding the plasterboard to the frame. Draw pencil lines to show where the timbers are.

2 With a trimming knife, cut away a section of plasterboard about 120mm square wherever your planned route crosses a frame member.

3 Chisel a groove in the exposed frame to hold the cable easily.

CABLE IN TRUNKING

Where it is not possible to lead cable under a floor or bury it in plaster, or if you want to reduce the labour, run the cable in surface-mounted mini-trunking. This protects the cable from damage as well as concealing it. The trunking is screwed or glued in place alongside mouldings; the plastic cover clips on once the cables have been put in place.

Cable run in mini-trunking does not have to be RCD protected (see page 57): it also provides a way of 'temporary' wiring until you are ready to re-decorate.

Feeding in cable from above

1 Cut out a square of the plasterboard to reveal the top of the timber frame and drill a hole through it large enough to take the cable. You can use a long drill bit and drill at a slight angle, or fit a right-angle adaptor and drill vertically.

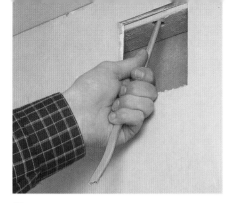

2 Thread the cable down through the drilled hole. Feed it down, ease it over each crosspiece of the frame and position it in the groove, allowing plenty of slack.

Feeding in cable from below

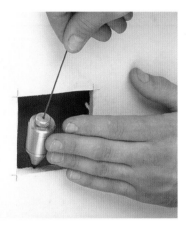

1 Cut out a square of plasterboard near floor level to reveal the base of the frame. Drill a hole through it for the cable, using a long bit and drilling at an angle or fitting a right-angle adaptor and drilling vertically. Feed a weighted cord down from the hole in the wall for the accessory.

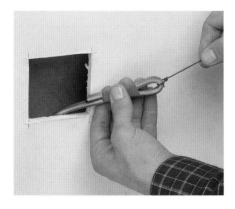

2 Tie the cord to the cable and draw it up carefully, easing it over each crosspiece of the frame, and position it in the prepared grooves. Leave plenty of slack in the cable.

Feeding in cable sideways

1 Push the cable along between the plasterboard panels. At each upright, draw out a loop of cable long enough to reach the next upright. Feed it along and set it in the prepared groove. When the cable is in the required position, cover the groove with a metal plate, pinned to the wood above and below, to prevent accidental damage from a nail or picture hook.

2 Cut new squares of plasterboard to replace the sections cut away. Tack them securely to the timbers at top and bottom, keeping the tacks well clear of the cable. Fill in round the edges of the squares with interior filler.

Fitting a plastic drylining box

Cut out the most convenient holes for the cable or cables to enter the mounting box. Hold the box in position, check with a spirit level that it is horizontal and draw a line round it. Cut away the marked section of plasterboard with a trimming knife. Feed the cable or cables into the box. Push the mounting box into the hole. Some boxes have spring-loaded fixing clips which simply snap into place as you push the box in; others have retractable lugs which you push out to hold the box. Prepare the conductors (page 39) and connect them to the terminals on the accessory faceplate. Screw the faceplate to the box.

Fitting a ceiling rose

A ceiling rose has to be screwed securely to a joist or to a batten between joists above the plasterboard of the ceiling. It is not sufficient to screw it through plasterboard into a cavity fixing device. Plasterboard is not strong enough to bear the weight of the light fitting and lampshade.

1 Mark the spot on the ceiling where you want the light fitting to hang.

2 Drill up through the ceiling at the marked spot. If the drill strikes a joist, probe with a bradawl about 50mm round the hole until you find space above the plasterboard. Insert a piece of spare cable through the hole to stick up above the ceiling as a marker.

3 Examine the ceiling from above, lifting a floorboard if necessary, and locate the hole with its piece of spare cable identifying it.

4 If the hole for the cables is between joists, cut a batten to length to fit between the joists.

5 Lay the batten between the joists with its edge touching the hole for the cables and mark the edge just beside the hole.

6 Drill through the centre of the batten, aligning the drill bit with the mark at the edge. The hole must be large enough to admit up to three or four cables.

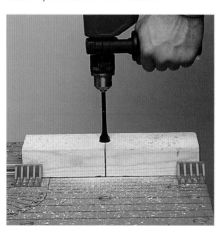

Alternatively If the hole for the cables is immediately below a joist, chisel out a groove in the bottom of the joist to fit the cables. You may need to nail a block of wood to the side of the joist to widen it so you can screw the ceiling rose exactly where you wish. Chisel a groove down the side of the block that will fit against the joist. The groove must be large enough to hold three or four cables easily.

7 Thread the cables through the hole in the batten and then through the hole in the ceiling.

8 Secure the batten to the joists with a nail hammered in at an angle at each end.

Alternatively Thread the cables through the prepared groove at the bottom of the joist and down through the hole in the ceiling. Fit the cables into the groove in the side of the prepared block of wood. Hold the block beside the joist and nail it in place, keeping the nails well clear of the cables.

9 Knock out the entry hole for the cables in the ceiling rose baseplate and thread the cables through the hole.

10 Hold the baseplate in place and insert a bradawl through the screw holes to penetrate through the plasterboard and pierce the timber above.

11 Screw the baseplate into place.

Making good

Do not connect any sockets or switches until the walls or ceilings are made good or the connections are likely to be dislodged as you make the necessary repairs.

1 Fill all the chases cut in the plaster with new plaster or interior filler and leave it to dry completely. For the best finish, fill them in two stages and sand the repair smooth when the filler has set hard.

2 Replace any floorboards you have lifted. If any board is likely to pinch a cable where it starts to run up a wall from below the floor, cut a notch in the end of the board before you replace it. Boards, or sections of boards, that you may need to lift again in the future – to add a junction box in a lighting circuit, for example – should be screwed down, not nailed. They can then be lifted without being damaged.

Adding a spur to a ring main circuit

An easy way to add a socket outlet to your wiring system is to run a spur from an existing outlet on the circuit.

• The spur is wired from the back of the outlet and can supply one new single, double or triple socket outlet, or one fused connection unit (FCU).
• Care is needed to find a suitable socket outlet for the spur connection. You must not use one that is already supplying a spur or is itself supplied as a spur.
• Make sure that the spur will not increase the floor area of the rooms served by the circuit to more than 100m².

Tools *Suitable tools for preparing the route (page 57): insulated screwdrivers; circuit tester; trimming knife; wire cutters and strippers; pliers.*

Materials *2.5mm² three-core cable; cable clips; green-and-yellow earth sleeving; mounting box with grommets and fixing screws; socket with switch.*

Finding a supply socket

1 Turn off the power at the consumer unit and take out the fuse or switch off the MCB protecting the circuit you want to work on. Check that the socket outlet is dead – for example, by plugging in a lamp that you know to be working.

2 Unscrew the faceplate of the socket outlet you plan to use for the spur. Ease it away from its mounting box until you can see the cables. If there is only one, the outlet is on a spur. If there are three, it is supplying a spur. Neither can be used to supply another spur.

3 A socket outlet with two cables may be suitable, but check with a circuit tester (page 64) before you go ahead. It could be the first outlet on a two-outlet spur installed earlier (but no longer permitted).

Preparation

When you have found a suitable socket outlet, ease its faceplate out and undo the terminal screws. Release the cores so that you can remove the faceplate. Remove the screws that hold the mounting box in place and carefully draw it out of its recess.
  Prepare the route (page 57) for the cable from the supply socket to the new socket. Feed the cable ends into the mounting boxes for the new and the supply sockets after fitting grommets in the knock-out cable entry holes.
  Fix the cable and boxes in place (page 57) and make good the wall damage. When the plaster has set, prepare the new cable ends for connection (page 39). Put green-and-yellow sleeving on the bare earth cores.

Connecting at the supply socket

1 Match the cores on the new spur cable with those on the existing circuit cables – brown to red, blue to black, and green-and-yellow to green-and-yellow.

All projects carrying this symbol are notifiable (see pages 6-7)

2 Connect the cores securely to the terminals on the faceplate. The red and brown cores go to the L terminal, the black and blue cores to the N terminal, and the green-and-yellow-sleeved earth cores to the E terminal (or ⏚).

USING A CONTINUITY TESTER TO CHECK A SPUR

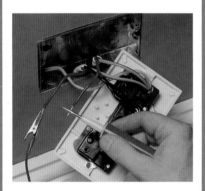

Use a continuity tester to check whether a socket outlet with two cables connected to it is on a ring main circuit.

1 Turn off the power at the consumer unit, remove the fuse or switch off the MCB protecting the circuit you will be working on, and unplug all appliances on the circuit.

2 At the socket outlet, disconnect the two red cores.

3 Attach the tester's clip to one end of each core and touch the probe to the other. If the socket is part of a ring circuit, the bulb on the tester will light up because you have completed the circuit; this is a suitable supply outlet for a spur.
   If the tester does not light up, the outlet is itself on a spur and cannot be used to supply one.

3 Fold the cables carefully into the mounting box and position the faceplate over the box.

4 Screw the faceplate to the mounting box. Do not overtighten the screws or you may crack the faceplate.

Connecting at the new socket

1 Connect the new cable cores to the terminals – brown at L, blue at N and green-and-yellow at E (or ⏚). Plus a flying earth link (see page 52) for metal boxes.

2 Fold the cables carefully into the mounting box and position the faceplate over the box. Screw it in place.

3 At the consumer unit, replace the circuit fuse or switch on the MCB and restore the power supply.

⊛ Fitting a fused connection unit

A fused connection unit (FCU) is used to connect fixed appliances to the power supply.

• An FCU is the only kind of power supply point (except for a shaver socket) that is allowed in most bathrooms. It can supply an extractor fan or a wall-mounted heater, both of which must be out of reach of a person using the bath or shower (see page 34 for more details).
• The FCU can be installed as part of a new ring circuit, as a replacement for one of the socket outlets on an existing circuit, or can be installed on a spur (see page 63).
• A cartridge fuse is housed in a small fuseholder in the FCU faceplate. The fuseholder can be removed by undoing a screw or can be prised open with a screwdriver to give access to the fuse. This should be a 3amp fuse for appliances rated up to 700 watts and a 13amp one for higher-wattage appliances.
• All FCUs fit a standard single mounting box. They may be switched or unswitched, and may have a neon indicator light. An unswitched FCU is suitable only for an appliance that has its own on/off switch. Do not use a switched FCU in a bathroom.

• For most appliances, choose an FCU that has a flex outlet at the front or side. In cases where the wiring from the FCU goes to a flex-outlet plate (see page 53), you do not need an entry hole: the wiring from the FCU to the flex outlet plate is done in two-core-and-earth cable, which can be buried in plaster or otherwise concealed.

Tools *Trimming knife; wire cutters and strippers; pliers; insulated screwdriver. Perhaps tools for preparing the route (page 57).*

Materials *Appliance in place with flex fitted; appropriate fuse.*

1 Turn off the power at the consumer unit and take out the fuse or switch off the MCB protecting the circuit you will be working on.

2 If the FCU is replacing an existing socket outlet, unscrew the faceplate and disconnect the cable cores from their terminals.

3 If the FCU is on a new spur, install the spur (page 63) up to the point where the mounting box and spur cable are in place. Remember that for many of the appliances connected to an FCU, the FCU should be above worktop height so that the switch and fuse are easily accessible.

4 Prepare the ends of the cable and the flex for connection (page 39). Remember to sleeve the bare earth conductor of the cable.

5 Feed the flex from the appliance to the back of the FCU. If there is a flex grip, release it sufficiently to let the flex through and then tighten it securely.

6 Connect the brown (live) core of the appliance flex into the terminal marked L and Load (or Out).

7 Connect the blue (neutral) core to the terminal marked N and Load (or Out)

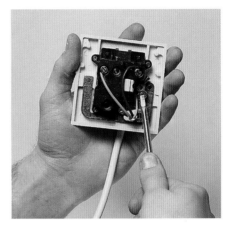

8 Connect the green-and-yellow earth core to one of the terminals marked E (or ⏚). If the FCU has only one earth terminal, do not connect the core yet. It will share the earth terminal with the spur cable earth core and you should insert them both together when you are connecting the cable to the FCU.

Connecting a spur cable to the FCU

1 Connect the brown (live) spur cable core to the terminal marked L and Mains (or Feed or Supply or In).

2 Connect the blue (neutral) spur cable core to the terminal marked N and Mains (or Feed or Supply or In).

3 Connect the green-and-yellow-sleeved earth core of the spur cable to the second terminal marked E (or ⏚).

FIXED APPLIANCES

MAINTENANCE AND REPAIRS

Alternatively If there is only one earth terminal, connect the earth cores from both flex and cable to it.

4 Add a flying earth link between the earth terminals on the faceplate and in a metal mounting box (see page 52).

5 Fit the correct cartridge fuse into the fuseholder in the FCU faceplate.

6 Fold the cable and flex neatly into the mounting box and press the faceplate against it.

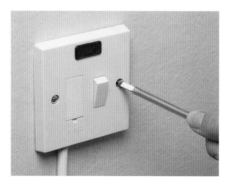

7 Screw the faceplate into place. Do not over-tighten screws or the plastic may crack.

CONNECTING AN FCU AS PART OF A RING CIRCUIT

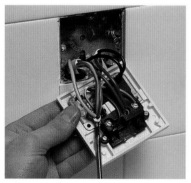

Connect the flex cores as for an FCU on a spur.

Connect the existing circuit cable cores into the terminals marked Mains (or Feed or Supply or In) as for an FCU on a spur – red to L, black to N, green-and-yellow to E (or ⏚). There will be two conductors to screw into each terminal.

Screw the FCU to the mounting box as for one on a spur.

Wiring electrical appliances in a bathroom

For safety, any electrical appliances installed in a bathroom should be run from a fused connection unit outside the room. Follow these instructions for preparing the wiring for an electrically heated towel rail or oil-filled radiator, extractor fan or a shaver socket.

For a towel rail, oil-filled radiator, or extractor fan, cable runs from the back of the flex-outlet plate (page 53) to a switched fused connection unit (FCU) outside the bathroom. The FCU is fitted on a spur led from a socket outlet on the ring main circuit, as shown.

Tools *Suitable tools for preparing the route (page 57); trimming knife; pliers; insulated screwdrivers; wire cutters and strippers.*

Materials *Two flush single mounting boxes; 2.5mm² two-core-and-earth cable, cable clips; green-and-yellow sleeving for the earth cores; FCU; 13amp cartridge fuse; flex-outlet plate; towel rail, radiator, or extractor fan (or shaver socket).*

Preparation

1 Check the earthing and bonding (pages 31-32) and remedy any defects. All the metallic parts in the room must be cross-bonded to one another and to earth.

2 Turn off the power at the consumer unit and remove the fuse or switch off the MCB protecting the circuit you want to work on.

3 Find a suitable supply socket outlet from which to run the spur (page 63). Remove the faceplate and undo the terminal screws to release the conductors.

4 Prepare the route (page 57). Lead the spur cable from the supply socket outlet to the wall outside the bathroom where the FCU will be sited, and from there to the bathroom wall where the flex outlet plate is to be fixed.

FIXED APPLIANCES

Wiring outside the bathroom

Cable from fused connection unit to flex outlet plate or shaver socket in bathroom

Switched fused connection unit, flush fitted on a spur from the ring circuit

Spur cable to fused connection unit from ring circuit, chased into wall

Supply socket on ring circuit for spur cable

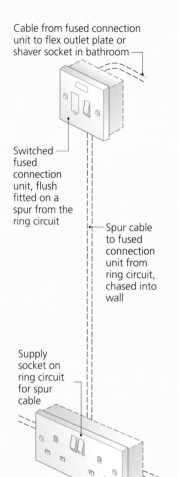

Cable to flex outlet plate

Spur cable from supply socket outlet

Connecting at the FCU

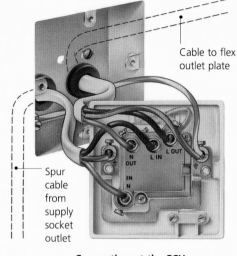

Spur cable to FCU

Ring circuit cables

Connecting at the supply socket

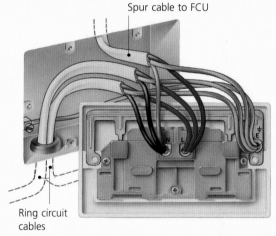

5 Prepare recesses for the mounting boxes for the FCU and the flex outlet plate.

6 Fix two lengths of cable in place, one from the supply socket to the FCU, the other from the FCU to the flex outlet plate.

7 Feed the ends of the cables into the mounting boxes after fitting grommets in the knock-out holes. Screw the boxes into place. Feed the end of the appliance flex through to the back of the flex outlet plate. Prepare the new cable and the flex cores for connection (page 39).

8 Repair the plaster and wait for it to dry. Replace any floorboards.

Connecting at the socket outlet

1 Connect the cores of the new spur cable – brown with red, blue with black and green-and-yellow with green-and-yellow.

2 Add a flying earth link between the earth terminals on the faceplate and in a metal mounting box (see page 52).

3 Press the socket onto the mounting box without disturbing the conductors and screw it in place.

17th EDITION RULES

There are new rules about bathroom wiring – see box on page 34

All projects carrying this symbol are notifiable (see pages 6-7)

Connecting at the FCU

1 Connect the spur cable from the supply socket to the FCU. The brown (live) core goes to the terminal marked L and Mains (or Supply, Feed or In). The blue (neutral) core goes to N and Mains (or Supply, Feed or In). The green-and-yellow earth core goes to the nearer of the two terminals marked E or ⏚. If the FCU has only one earth terminal, do not connect it yet.

2 Connect the cable leading from the FCU to the flex outlet plate. The brown (live) core goes to the terminal marked L and Load (or Out), the blue core to N and Load (or Out), and the green-and-yellow-sleeved earth core to the second terminal marked E or ⏚. If there is only one earth terminal, connect both earth cores to it.

3 Add a flying earth link between the earth terminals on the faceplate and in a metal mounting box (see page 52).

4 Fit the 13amp fuse in the fuseholder in the FCU faceplate and screw the faceplate to the mounting box.

5 Then connect the cable that runs from the FCU to the flex outlet plate or shaver socket in the bathroom (opposite).

Connecting a heated towel rail, electric radiator or extractor fan

The flex for an electrical appliance in a bathroom is normally wired into a flex outlet plate that does not have a switch.

1 Run a cable from a fused connection unit (FCU) outside the bathroom (page 66) to the position of the flex outlet plate. Fit a flush mounting box into the wall and feed the cable into it then connect the new cable at the FCU and the flex outlet plate.

2 Connect the flex cores to one set of terminals – brown to L, blue to N, and green-and-yellow to E or ⏚.

3 Connect the cable cores in the same way to the other set of terminals – brown to L, blue to N and green-and-yellow to E or ⏚.

4 Add a flying earth link between the earth terminals on the faceplate and in a metal mounting box (see page 52).

5 Screw the faceplate to the mounting box.

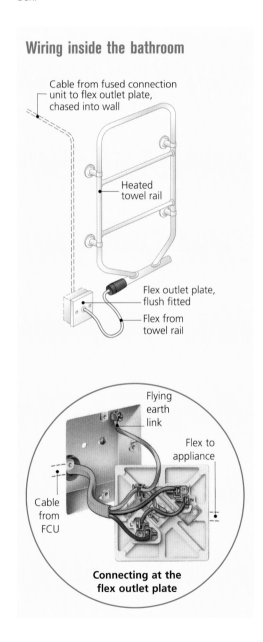

Wiring inside the bathroom

Cable from fused connection unit to flex outlet plate, chased into wall

Heated towel rail

Flex outlet plate, flush fitted

Flex from towel rail

Flying earth link

Flex to appliance

Cable from FCU

Connecting at the flex outlet plate

✋ Fitting a shaver point

The only electric socket outlet permitted in most bathrooms is a shaver point. Connect it to the main power supply via an FCU outside the bathroom (page 66).

• There are purpose-made mounting boxes for shaver supply units. Surface-mounted boxes are available but it is safer to fit a flush-mounting metal or (for plasterboard) a plastic drylining box into the wall.
• You can install the unit on a spur from an existing socket outlet on a ring main circuit.
• The preparation and general principles of installing a shaver supply unit on a spur from a ring-main circuit are the same as those described for connecting heated towel rails or oil-filled radiators at the supply socket or FCU (see opposite). In the bathroom, connect the cable run from the FCU to the shaver socket below).

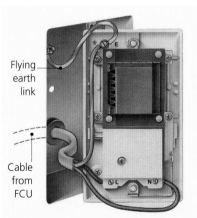

Flying earth link

Cable from FCU

Connecting at the shaver point

Installing a shaver point on a lighting spur

It is also common to install a shaver socket on a spur led from a junction box inserted in the lighting circuit above the bathroom (see page 91).

1 Lead the spur cable from the lighting circuit to the shaver supply unit position, using 1mm² twin-core-and-earth cable.

2 Fit the mounting box for the shaver supply unit at about shoulder height on the wall. Feed in the cable through a grommet.

3 Prepare the end of the cable and connect the cores into the terminals on the shaver supply unit faceplate – brown to L, blue to N, and green-and-yellow to E or ⏚.

4 Fold the cable neatly into the mounting box and screw on the faceplate. Restore the power supply to the lighting circuit.

✋ Installing an electric shower

An electric shower uses a lot of power in short bursts. It must have its own circuit leading from the consumer unit, protected by an RCD.

• For electric showers up to 7.2kW, the circuit can be run in 6mm² two-core-and-earth cable protected by a 32A MCB; up to 9.2kW, you will need 10mm² cable and a 40A MCB; up to 10kW, you would need to go to a 50A MCB.
• Under the latest Wiring Regulations (see Box on page 35), the shower circuit must be RCD protected, which means either using a spare RCD-protected 32A or 40A MCB in the exising consumer unit or replacing a spare MCB with an RCBO (see page 72) or fitting an extra enclosure as described on page 55.
• The circuit cable runs first to a pull-cord switch with neon and mechanical indicators.
• From the switch, the cable runs to the back or base of the shower heater unit, where it is connected to a terminal connector block inside the unit.
• Follow the maker's instructions on fitting the unit to the wall and plumbing it in.

Tools *Suitable tools for preparing the route (page 57); insulated screwdrivers; trimming knife; wire strippers and cutters; pliers.*

Materials *Two-core-and-earth cable of appropriate size; green-and-yellow plastic sleeving for earth conductors; cable clips; 45amp double-pole pull-cord switch with neon warning light and mechanical on/off indicator; shower heater unit.*

All projects carrying this symbol are notifiable (see pages 6-7)

Preparation

1 Plumb the unit over the bath or in a shower cubicle.

2 Take off the front cover of the shower heater and hold the back plate against the wall. Mark the wall through the cable entry point so that you know where to lead the cable, and also through the screw fixing points. Drill and plug the screw fixing points.

3 Check the main and supplementary bonding (pages 32-33).

4 Prepare the route for the cable (page 57). There is no need to switch off the electricity until you are ready to make the final connection at the consumer unit. Make the route from the consumer unit to the pull-cord switch position on the ceiling of the bathroom or shower room. Continue the route from the switch to a point on the ceiling above the shower and then vertically down the wall to the position of the shower heater unit.

5 Fit the baseplate for the switch to the ceiling following the steps for fitting a ceiling rose (page 62). Make sure that the fixing is secure enough to withstand the switch cord being pulled. Before you screw it in place, knock out two of the stamped circles for the cables to enter.

6 Lay the cable along the route, from the consumer unit to the double-pole switch and from the switch to the cable entry point on the shower heater unit. Allow sufficient cable at each end to reach all the terminals comfortably.

7 Feed the cable ends into the switch. Then feed the cable through the entry hole of the shower heater. Screw the unit to the wall.

8 Prepare all the cable ends for connection (page 39). Fit green-and-yellow sleeving on all the bare earth conductors.

9 If there is no spare suitable MCB on the consumer unit, fit a separate enclosure as described on page 55. You can get special 'shower protection' units that contain a double-pole 40amp RCBO or an (63amp/30mA) RCD feeding a 50amp MCB.

10 Complete the plumbing connections.

EARTH SAFETY

Cross-bond the shower supply pipework to earth by fitting an earth clamp to the pipe and connecting 4mm² single-core earth cable to it. Run this cable to the earth terminal on either the ceiling-mounted switch or the shower unit, whichever is the more convenient.

Connecting the ceiling switch

1 Insert the two green-and-yellow sleeved earth cores into the terminal marked E or ⏚ and connect them tightly in place.

2 Connect the brown and blue cores of the cable from the ceiling switch to the shower unit. The brown core goes to the terminal marked L and Load (or Out) and the blue core to the terminal marked N and Load (or Out). Check the neon indicator cores at the same time, one (it does not matter which) to the same terminal as the brown cable core and the other to the same as the blue core. Take care not to dislodge the bulb.

3 Connect the brown and blue cores of the cable from the consumer unit to the other terminals (marked Supply, Feed or In). Connect the brown core to the terminal marked L and the blue core to the terminal marked N. Fit the switch cover over the baseplate and screw the cover in place.

Connecting at the heater unit

Do not disturb the internal wiring already fitted in the shower unit.

1 Release the screws of the cable clamp, lead the cable under the clamp and screw the clamp back in place.

2 Connect the cores of the cable to the terminal connector block incorporated in the shower heater unit. Flex cores will already be connected at one side of the block. Connect the brown cable core to the terminal opposite the brown flex core, and the blue cable core to the terminal opposite the blue flex core.

Continued on page 72

Wiring for an electric shower

Connecting at the ceiling switch

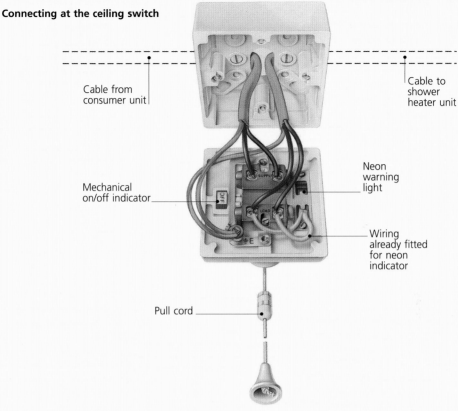

Cable from consumer unit

Cable to shower heater unit

Mechanical on/off indicator

Neon warning light

Wiring already fitted for neon indicator

Pull cord

Connecting at the shower heater unit

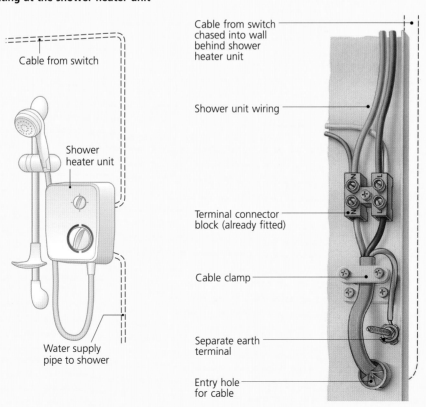

Cable from switch

Shower heater unit

Water supply pipe to shower

Cable from switch chased into wall behind shower heater unit

Shower unit wiring

Terminal connector block (already fitted)

Cable clamp

Separate earth terminal

Entry hole for cable

3 Instead of a pair of earth terminals on the connector block, there may be a single separate earth terminal marked E or ⏚. Connect the green-and-yellow sleeved cable earth core to this terminal.

Alternatively If the connector block has a third terminal for the earth core, connect the sleeved core to it opposite the green-and-yellow flex core already connected there.

Connecting the circuit at the consumer unit

1 Turn off the power at the consumer unit and switch off the RCD-protected MCB you will be using for the new circuit.

2 Unscrew the cover of the consumer unit. If the unit has a wooden frame, drill a hole through it and feed in the new cable.

Alternatively On a metal unit, knock out the entry hole immediately above the MCB you will be using. Fit a rubber grommet into the hole and feed in the cable.

3 Check that you have stripped off enough outer sheath from the cable for the cores to reach their terminals easily.

4 Connect the cores tightly to their terminals: brown to the terminal at the top of the MCB; blue to a free terminal on the neutral terminal block; sleeved green-and-yellow earth core to a free terminal on the earth terminal block.

5 Replace the cover of the consumer unit. If you have fitted a separate enclosure to provide RCD protection, ask your electricity supplier to connect this up.

RCBO PROTECTION

An RCBO (residual current circuit breaker with overload protection) combines the functions of an RCD and an MCB. It can be fitted in place of an MCB in most consumer units to provide RCD protection – in order to meet the 17th Edition Wiring Regulations.

RCBOs, however, are expensive and if you need more than around four it might be better to go for a complete new 17th Edition consumer unit (see page 55).

Installing a new cooker circuit

Plan the cable route (page 57) from the cooker switch back to the consumer unit. To meet Wiring Regulations, a new cooker circuit must be RCD-protected. If there is not an RCD-protected MCB to connect to, either fit an RCBO as for a shower (page 69) or fit a new enclosure (page 55).

Tools *Suitable tools for preparing the route (page 57); trimming knife; pliers; insulated screwdrivers; wire cutters and strippers.*

Materials *45amp cooker switch with mounting box; cooker connection unit(s) with mounting box(es); grommets; fixing screws and wallplugs; two-core-and-earth cable and green-and-yellow plastic sleeving of the correct size; cable clips; free-standing cooker or separate oven and hob; available MCB or RCBO of the correct rating.*

Putting in the cables

1 Prepare the route (page 57), leading it from near the consumer unit to the cooker switch. From there lead it to the cooker connection unit behind a free-standing cooker, or to connection units behind a separate oven and hob.

2 Fit the mounting box for the cooker switch and fit the mounting box(es) for the connection unit(s). Remember to remove knock-outs from the mounting boxes for the cables to enter, and to fit grommets in the holes.

3 Fit lengths of cable along the route from the consumer unit to the switch, and from the switch to the connection unit(s). Do not feed the cable into the consumer unit but feed the other cable ends into the mounting boxes. Allow enough spare cable at the ends to reach all the terminals easily. Prepare the cable ends for connection (page 39), remembering to sleeve the earth conductors.

4 Repair the plaster and wait for it to dry. Replace any floorboards you have lifted.

Continued on page 74

Understanding an electric cooker circuit

An electric cooker uses so much electricity that it must have its own circuit. If it shared a circuit with other appliances the circuit would frequently be overloaded.

At the consumer unit The cooker circuit is protected by a 30amp fuse or 32amp MCB for a cooker rated at up to 12kW.
• A 45amp fuse or 40amp MCB is needed for a cooker rated at above 12kW.
• A separate oven and hob operating on the same circuit may have a higher rating than 12kW.
• A circuit protected by a 30amp fuse or a 32amp MCB needs to be run in 6mm² two-core-and-earth cable; a circuit protected by a 45amp fuse or 40amp MCB needs larger 10mm² two-core-and-earth cable.

The cable runs from the consumer unit to a cooker switch This is mounted above worktop height on the kitchen wall – beside the cooker, not above it.

The cooker switch
This is a double-pole 45amp switch that disconnects both the live and neutral conductors. The unit may have a neon light that glows when the unit is switched on.
   Many older units included a 13amp socket outlet, but these are best avoided because of the possibility of trailing flexes being scorched by one of the cooker hotplates. If the control unit includes a socket outlet, a 30amp fuse or 32amp MCB will still be suitable for a cooker rated at up to 10kW.

Cooker switch

Consumer unit

Connection unit

Free standing cooker

For a free-standing cooker A second length of cable is buried in the wall to run from the cooker switch to a cooker connection unit fitted on the wall about 600mm above floor level behind the cooker. A 2m length of cable (not flex) runs from the connection unit to the cooker, so that the cooker can be drawn away from the wall when necessary. Use the same size of cable for this as for the rest of the circuit.

For a separate oven and hob These can be connected to one cooker switch if neither the hob nor the oven is more than 2m away from the switch.
• Make the connection by running two cables from the switch, one to the oven and the other to the hob.
• Alternatively, run one length of cable from the switch to one part of the cooker and a second length of cable from there to the second part of the cooker.

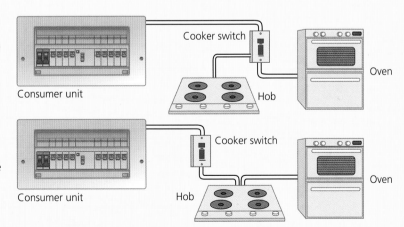

Cooker switch

Consumer unit

Hob

Oven

Cooker switch

Consumer unit

Hob

Oven

• It is possible to connect the cable directly into the oven and the hob (as they are permanently built-in, but it is much better practice to install a cooker connection unit (or two cooker connection units) as for a free-standing cooker.
• If either oven or hob would be more than 2m away from a single cooker switch, you will need two switches, but both can be connected to the same cooker circuit.

FIXED APPLIANCES

Connecting at the switch

1 Connect the cable from the consumer unit to the terminals of the cooker switch. Connect the brown core to the terminal marked L and In. Connect the blue core to the terminal marked N and In. Connect the green-and-yellow-sleeved earth core to the nearer of the terminals marked E or ⏚.

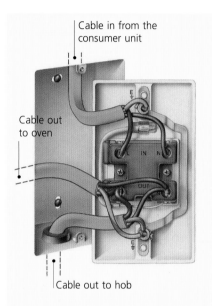

Cable in from the consumer unit

Cable out to oven

Cable out to hob

Connecting at the cooker switch
Two cables lead to a separate oven and hob. Only one goes to a free-standing cooker or to the oven and hob if they are to one side of the control unit.

2 Connect the cable leading out to the connection unit(s) to the terminals behind the switch plate. Take the brown core to the terminal marked L and Out, and the blue core to the terminal marked N and Out. Connect the green-and-yellow-sleeved earth core to the nearer of the terminals marked E or ⏚.

If you are leading separate cables to an oven and hob, there will be two outgoing sets of cores. Match the cores in pairs – brown with brown, blue with blue and green-and-yellow with green-and-yellow. Insert the pairs of cores into the correct terminals and screw them in place.

Connecting a free-standing cooker at the connection unit

1 Remove the screws holding the cover to the metal frame. Then unscrew and remove the cable clamp at the bottom of the frame.

Wiring at the connection unit

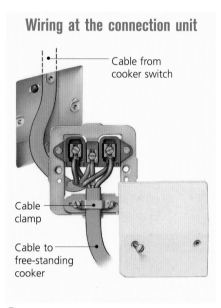

Cable from cooker switch

Cable clamp

Cable to free-standing cooker

2 Pair the cable cores from the switch and to the cooker – brown with brown, blue with blue, and green-and-yellow with green-and-yellow. Screw the pairs into the terminal block on the frame – brown to L, blue to N and green-and-yellow to E or ⏚.

3 Screw the frame to the mounting box and screw on the cable clamp.

4 Screw the cover of the connection unit in place.

Connecting to a free-standing cooker

1 Remove the metal plate covering the terminals on the back of the cooker. Release the cable clamp.

2 Connect the supply cable cores to their terminals – brown to L and blue to N. Sometimes the cores have to be bent round a pillar and held down with brass washers and nuts. Make sure enough insulation has been removed for the bare wire to wind round the pillars. Connect the green-and-yellow-sleeved earth core to E or ⏚.

3 Screw the clamp over the cable. Screw the plate back over the terminals.

Connecting to the consumer unit

1 Turn off the main switch at the consumer unit and switch off the MCB for the cooker circuit. Make sure the cooker switch and all cooker controls are off. Replace MCB with RCBO (page 72) if necessary.

2 Remove the screws of the consumer unit cover and take it off.

3 Drill through the frame of the consumer unit if it is wooden, or knock out an entry hole and fit a grommet if it is a metal or

plastic one. Feed in the cable. Make sure that you have removed enough outer sheath for the cores to reach the terminals.

4 Connect the blue core to a spare terminal at the neutral terminal block and connect the brown core to the terminal on the spare MCB or RCBO.

5 Connect the green-and-yellow-sleeved earth core to a spare terminal at the earth terminal block.

6 Switch on the MCB or RCBO.

7 Screw on the cover of the consumer unit and turn the main switch back on.

CONNECTING A SEPARATE OVEN AND HOB

If you have led two cables from the control unit, one each for the oven and hob, connect each cable in the same way as a free-standing cooker (left) and secure under the clamp.

Alternatively, if you have led one cable from the cooker switch to the first component (via a connection unit), you will have two cables to connect and clamp there – one from the cooker switch and one to the second component.

Match the cores – brown with brown, blue with blue, and green-and-yellow with green-and-yellow. Connect them to the terminals – brown to L, blue to N, and green-and-yellow to E or ⏚.

Connect the ongoing cable to the second component as for a free-standing cooker.

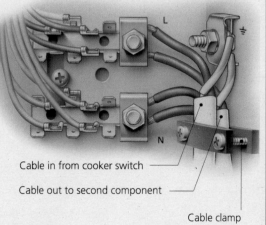

Cable in from cooker switch

Cable out to second component

Cable clamp

Installing a new ring main circuit

A ring main circuit to supply socket outlets starts at the consumer unit. A 2.5mm² two-core-and-earth cable runs round the rooms supplied by the circuit, looping into and out of each socket outlet before returning to the consumer unit to complete the ring. All new socket outlet circuits must now be provided with RCD protection.

If you have a modern consumer unit, fitted with a 30mA RCD and with a spare RCD-protected 32amp MCB, and are installing a new ring circuit – perhaps to serve an extension – you can connect the new circuit to the consumer unit. If your consumer unit has spare non RCD-protected MCBs, you may be able to replace one with a 32A RCBO (page 72) as for an electric shower circuit (page 69). For older consumer units, fit a new enclosure (page 55) or have a new 17th Edition consumer unit fitted (page 55).
• Any number of single, double or triple socket outlets can be installed on the new ring main circuit, but the floor area of the rooms served by the circuit must not exceed 100m².
• If a socket outlet is needed in a position that is not on the most convenient route for the ring circuit, you can run a non-fused spur to it from one of the socket outlets on the ring, but this will reduce the possibility of adding spurs later.
• The number of non-fused spurs must not exceed the number of outlets on the ring.

Tools *Suitable tools for preparing the route (page 57); trimming knife; insulated screwdrivers; wire strippers and cutters; pliers.*

Materials *Mounting boxes; screws and wallplugs; grommets; socket outlets (and perhaps FCUs); 2.5mm² two-core-and-earth cable; green-and-yellow plastic sleeving for bare earth cores; cable clips; fuses for FCUs; available 32amp MCB or RCBO in the consumer unit.*

Preparation

1 Prepare the route (page 57), leading it from the consumer unit to the nearest socket outlet position on the new ring main circuit, and on from there to each socket outlet position in turn. Prepare the route for any spurs branching off the ring main circuit. Lead the route back from the final socket outlet position to the consumer unit.

2 Prepare recesses for the mounting boxes of the new socket outlets.

3 Fit the mounting boxes in position and knock out the most convenient cable entry holes. Fit grommets in metal boxes.

4 Lay the cable along the route (page 58). Start above the consumer unit, leaving plenty of spare cable for the cores to reach the terminals in the unit easily. Do not interfere with the consumer unit. Go along the route, taking the cable in and out of the most convenient entry holes at each mounting box. Cut the cable at each box, leaving 100mm spare at each cut. Fit branch cables along the route to any spurs. Leave 100mm spare at each end of the cable. Lead the cable from the last socket outlet on the circuit to the consumer unit.

5 Prepare all the cable ends (page 39) for connection at the socket outlets.

6 Repair the plaster and let it dry. Replace any lifted floorboards. Do not fix them down if you plan to remove old wiring later.

Connecting the socket outlets

1 At each socket outlet, pair the two sets of cable cores – brown with brown, blue with blue, and green-and-yellow with green-and-yellow.

2 Connect the cores to the terminals on the rear of the faceplate – browns to L, blues to N, and green-and-yellows to E or ⏚. Add a flying earth link (see step 3 on page 52) between the earth terminal on the faceplate and the earth terminal in the mounting box if the latter is metal.

3 Fold the two cables neatly back into the mounting box and press the faceplate into position over the box. Screw it in place until there is no gap between the faceplate and the mounting box. Do not overtighten the screws or the plastic may crack.

Continued on page 78

Planning a ring circuit

1 Draw a plan of the rooms that the ring circuit will supply. The new circuit might serve the kitchen only, all the other ground floor rooms, all the upper floor rooms or the socket outlets in an extension.

2 Mark on the plan where socket outlets are needed – and where they might be needed in the future. Most rooms will need sockets on at least two sides. Large rooms may need them on three sides and at two places along the sides. A choice of socket positions does away with flexes trailing across rooms.

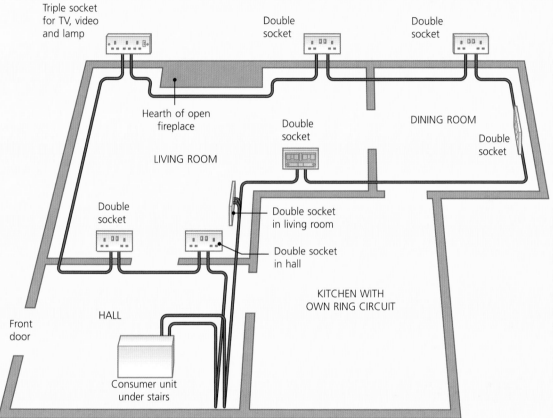

Triple socket for TV, video and lamp

Double socket

Double socket

Hearth of open fireplace

LIVING ROOM

Double socket

DINING ROOM

Double socket

Double socket in living room

Double socket in hall

Double socket

KITCHEN WITH OWN RING CIRCUIT

Front door

HALL

Consumer unit under stairs

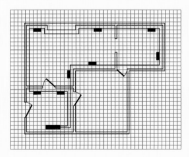

Drawing a plan of your rooms to scale on graph paper (left) will help you to work out the most practicable route for the cable and the positions that will be convenient for installing the sockets.

3 Install a double socket at each new outlet. In a kitchen or living room, triple sockets may be wiser. At some positions you may want to fit a fused connection unit (FCU) rather than a socket for a built-in appliance, such as an extractor fan or cooker hood (page 64).

4 The recommended number of double socket outlets is: kitchen and living room 6 to 10 (each); bedrooms 4 to 6 (each); dining room 3; hall, garage and utility room 2; study/home office 6; landing/stairs and loft 1.

Connecting spur socket outlets

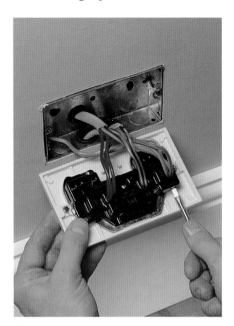

1 At socket outlets supplying spurs, there will be three sets of cable cores – one from the previous outlet on the circuit, one going to the next outlet, and one for the spur. Match the cores in threes, then follow steps 2 and 3 from the previous section.

2 The socket outlet at the end of a spur has only one set of cores to connect. Follow steps 2 and 3 from the previous section.

Connecting FCUs

The appliance should be fitted in place, complete with its flex.

1 Feed the flex through the entry hole to the back of the fused connection unit (FCU) and prepare the end of the flex for connection (page 39).

2 Connect the flex cores – brown to L and Load (or Out), blue to N and Load (or Out), and green-and-yellow to the nearer of the two terminals marked E or ⏚.

3 Pair the cores from the two cables, brown with brown, blue with blue, and green-and-yellow with green-and-yellow. (If the FCU has been fitted at the end of a spur there will be only one cable.)

4 Connect the cable cores to their terminals – brown cores to the one marked L and Mains (or Supply or Feed or In), blue cores to the one marked N and Mains (or Supply or Feed or In) and green-and-yellow cores to the nearer of the two terminals marked E or ⏚. Add a flying earth link to the earth terminal in a metal box (page 52).

5 Fit a 3amp or 13amp cartridge fuse in the fuseholder to suit the wattage rating of the appliance.

6 Fold the cable(s) and flex neatly into the mounting box and press the FCU faceplate into position. Screw it to the mounting box.

Having a consumer unit installed

If you do not have a modern consumer unit, arrange with your electricity supply company (or another approved electrical contractor) to install one. The installer will connect it to the meter and reconnect the new and existing circuit cables to it.

A 17th Edition consumer unit (page 55) can provide RCD protection on all circuits, but will also allow for at least one non-RCD circuit, with cable surface-run in mini-trunking. You might want this to supply a freezer so that food is not ruined should an RCD trip while you are on holiday.

Testing the new circuit

Test the continuity of each conductor in the circuit (see box, far right). Check and tighten all connections, if necessary.

Connecting at the consumer unit

1 Turn off the main switch on the consumer unit. Remove the retaining screws and the cover of the consumer unit.

SAFETY WARNING

The main switch disconnects only the MCBs and the cables leading out from the consumer unit to the household circuits. It does not disconnect the cables entering via the meter from the service cable. Do not interfere with these cables. They are always live at mains voltage.

2 Drill though the top of the consumer unit or knock out an entry hole and fit a grommet. Feed in the cables and prepare the cable ends. Make sure that enough sheath has been removed for the cores to reach the terminals easily.

3 Match the cores of the circuit cables – brown with brown, blue with blue, and green-and-yellow with green-and-yellow. Prepare core ends for connection.

4 Connect the cores to their terminals – brown cores to the terminal on the spare 32amp MCB, blue cores to the neutral terminal block, and green-and-yellow-sleeved earth cores to the earth terminal.

5 Switch on the MCB. Replace the cover of the consumer unit.

6 Turn on the main switch and test the circuit by connecting a table lamp you know to be working to each socket outlet in turn or use a plug-in socket tester.

Disconnecting the old wiring

1 At the fuse board, turn off the main switch or switches that controlled the superseded circuits and remove any fuses.

You can arrange for the electricity board to disconnect the old system from their meter.

2 Trace the cables leading from the switch to the old socket outlets down to the floor or up to the ceiling. Cut them off there.

3 At each of the old socket outlets, unscrew the faceplate and release the cable cores from the terminals. Unscrew and prise out the mounting box.

4 To remove all the old wiring you may have to lift several floorboards. Remove as much as you can and wherever possible any junction boxes, but leave any conduits in the wall. Repair the walls as necessary, and replace any floorboards you lifted.

5 Remove the old main switch and old fuse box once they have been disconnected by the supply company. Remove the cables that lead from them. If the old system has not been disconnected, do not remove the switch or fuse box. Keep the switch off and the fuse out.

TESTING A NEW RING CIRCUIT

When you install a new ring circuit, use a circuit tester to check that all the connections are sound *before* you connect it at the consumer unit.

Where the two cable ends approach the consumer unit, clip the tester to the end of one brown core and put the probe on the other. If the tester fails to light, tighten the connections all round the circuit. Carry out the same checking procedure with the blue and earth cores.

If the tester lights when it is linking two *different* cores there is a serious wiring fault and you should call in a qualified electrician.

Choosing light switches

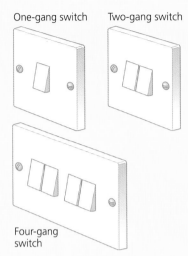

One-gang switch   Two-gang switch

Four-gang switch

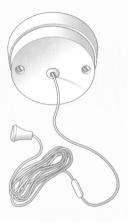

Plastic switches The most common are white plastic, designed to blend with light-coloured paintwork. There is no earth terminal for the cable behind plastic light switches; it is in the mounting box.

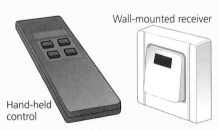

Three-gang switch   Four-gang switch

Metal switches Metal switches are made in satin finish or shiny brass, chrome or stainless steel; some brass switches have decorative scrolling at the edge. All metal switches must have an earth terminal on the rear of the faceplate. The switch rockers are fixed in plastic inserts.

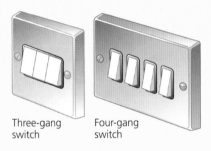

Hand-held control

Wall-mounted receiver

Remote control switches Remote control switches are operated by an infra-red signal from a control unit; this triggers a wall-mounted receiver which acts as the switch. The receiver must be in sight of the control unit but can be up to 8m away. The switch can also be operated by hand.

Pull-cord switches

If a pull-cord switch is fitted in a bathroom the mechanism must be well out of reach of anyone who is using the bath or shower. These switches are also useful in bedrooms.

Two-gang dimmer

Dimmer switches Dimmer switches control the amount of light a lamp gives out. Dimmer switches can have push, sliding or rotary control; others have a touchplate. Ordinary dimmers cannot be used with fluorescent lights, but some can be used with low-voltage halogen lights. Some compact fluorescent lamps (CFLs – see page 84) are dimmable and there are special dimmer switches for LED (light emitting diode) light bulbs.

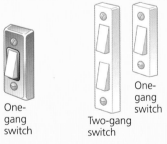

One-gang switch

Two-gang switch

One-gang switch

Architrave switches Architrave switches can be fitted where there is limited wall space – for example, where two door frames are so close together that there is not enough space between them for a standard switch. Mounting boxes are made to suit them (see right).

MOUNTING BOXES

Boxes for surface mounting can be metal or plastic, but flush-fitting boxes for solid walls are only metal. A single mounting box will take a faceplate with one, two or three switches (gangs); a double box is needed for a faceplate with four or six switches. All metal boxes must have an earth terminal for the earth core of the switch cable; many plastic boxes also have one.

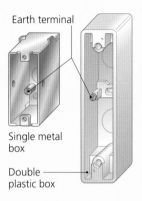

Earth terminal

Single metal box

Double plastic box

Earth terminal

25mm metal box

Boxes for architrave switches

Metal boxes for flush-mounted architrave switches are 30mm deep; plastic boxes for surface mounting are 16–18mm deep. Both types are available for mounting one-gang and two-gang switches.

Earth terminal

16mm metal box

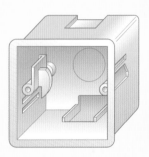

Plastic drylining box with rotating lugs

Boxes for stud partition walls
Plastic 'drylining' boxes are made for flush-fitting switches (and socket outlets) in plasterboard walls. They fit in a hole cut in the board, and have spring-loaded or rotating lugs at each side which grip the inner face of the board when the box is fully inserted. Single and double boxes are available.

Flush-mounting metal boxes A standard mounting box for a flush-fitted switch is made of metal and is 25mm deep. It is set in a recess in the wall and is screwed in place at the back. A plaster-depth box 16mm deep is also made and can be used for normal light switches. Deeper (35mm) flush-mounting boxes may be needed for some dimmer switches.

Surface-mounting boxes Surface-mounting boxes for light switches are normally plastic, but metal boxes are available to match metal light switches. Depths range from 16mm to 47mm; no recess is needed in the wall.

Earth terminal

16 mm plastic box

MAINTENANCE AND REPAIRS

Standard light fittings

A ceiling rose is the joining point for the flex of a pendant light and the cables of the lighting circuit. The terminals on the rose baseplate are hidden by the screw-on cover.

Ceiling rose

The body of a lampholder is concealed inside a two-part heat-resistant cover. The top part conceals the terminals to which the pendant flex is connected. The lampshade ring secures the lamp-shade. You can buy pendant sets with the ceiling rose and lampholder pre-wired.

Lampholder

Batten holder

A batten lampholder is a ceiling rose and lampholder combined. Some batten lampholders have a deep shield over the metal bulb holder; they are for use in bathrooms to prevent accidental contact with the metal. Angled batten lampholders carry the lampholder at an angle to the base, and are intended for wall mounting – in an under-stairs cupboard, for example.

Low-voltage fittings
For essential information about low voltage lighting, see page 95.

Track lighting
Mains-voltage track lighting can be wall or ceiling mounted and lengths of straight or curved track can be joined. Several types of small, neat light fittings are made to clip into the track. You can mount track over an existing lighting point, or lead flex to it from a lighting point a short distance away. Make sure that you will not overload the circuit if you fit a multi-spot track. If low-voltage halogen bulbs are used, the transformer can be built into the track, so no extra wiring is needed.

WALL LIGHT FITTINGS

Small conduit box (BESA box) to use with internal wall light

Most wall lights have a circular backing plate which can be secured direct to a small conduit box (known to electricians as a BESA box), recessed into the wall. The wiring is then done inside the box, using a terminal connector block cut from a strip of terminal connectors.

Terminal connector strip

The alternative is to use a fitting known as an LSC (luminaire supporting coupler) which comes in two parts: the 'socket' is attached to a conduit box in the wall; the 'plug' is on the light fitting.

Uplighters are secured direct to the wall; again, the wiring connection can be done in a conduit box recessed into the wall.

DECORATIVE LIGHT FITTINGS

There is a wide range of decorative light
fittings available that are designed to be fitted
flush to the ceiling surface or recessed into it.
Most are best suited to the junction box
method of wiring (see page 48), where there is
only a single cable leading to the light fitting.
• Light fittings with hollow baseplates use a
terminal connector block concealed within the
baseplate to which the circuit cable is
connected. Sometimes this block is supplied
already fitted; if not, you can cut your own
block from a terminal connector strip.
• Light fittings with a flush baseplate must be
installed over an enclosure (for example, a
conduit box) recessed into the ceiling to
contain the terminal connector block that joins
the fitting's flex to the circuit cable.
• Recessed light fittings are set in a hole cut in
the ceiling surface. Connections to the lighting
circuit cable are made within the ceiling void.
• Spotlights may be surface-fixed or recessed,
and direct light in one direction only. 'Eyeball'
lights are recessed and swivel to beam light
at an angle. Diameters vary but all need 100
to 125mm of clear space above the ceiling.
• Spotlights and recessed lights can use
reflector and halogen bulbs (page 84).
• Downlighters are fully or partly recessed, and
cast light downwards only.
• Most economical but least attractive is a
fluorescent tube, covered with a diffuser.
• Heavier lights, such as chandeliers, must be
supported by a chain secured via a hook to a
ceiling joist.

Heat-resisting cover

Recessed eyeball downlighter

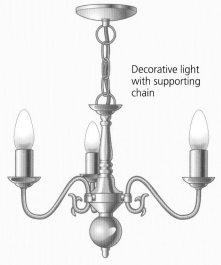

Decorative light with supporting chain

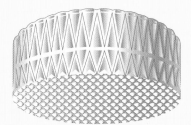

Flush ceiling light

Surface-mounted spotlight

Fluorescent light

CHOOSING LIGHTS FOR A BATHROOM

Strict regulations govern the use of electric light
fittings in wet areas. When choosing lights for a
bathroom, think about the position of the fitting and choose a light with an IP rating
(indicating how waterproof it is) suitable for the appropriate zone (see page 34). If water
jets are likely to be used for cleaning purposes in any of these zones, a fitting rated to a
minimum of IPX5 must be used.

Zone 1	IPX4, protected by an RCD if mains voltage (240V)
Zone 2	IPX4

Choosing light bulbs

The most usual source of lighting for the home, commonly called a light bulb, is more properly called a lamp. Up until 2010, most light bulbs were incandescent with a white-hot tungsten filament but this inefficient type of bulb has now largely been phased out, both in the UK and all across Europe, and replaced by low-energy (compact fluorescent and LED) and low-voltage lamps (see page 95).

• Most light bulbs have a bayonet cap (BC) fitting to push and twist into a BC lampholder. Some have Edison screw (ES) caps that screw into an ES lampholder. There are small versions of both – called SBC and SES.

• Halogen lamps give a brighter, whiter light than tungsten filament lamps.

• See page 76 for details of fluorescent tubes and compact fluorescent lamps.

Pearl (BC)

Clear (BC)　　　　White (BC)

Reflector bulbs

Used in spotlights and recessed lights, reflector bulbs direct light in a wide cone. They come in three main sizes: R50, R63 and R80, usually with an ES fitting.

Halogen spots

Some mains voltage halogen spots can replace normal reflector bulbs; others need their own fitting with a 'twist-and-lock' action. For more information on low-voltage lighting, see page 95.

Standard light bulbs The light bulbs known as General Lighting Service (GLS) bulbs are very familiar (and are illustrated above), but now only available in low-wattage versions. They have been used in pendant lights and table and standard lamps, but are now being replaced by low-energy or low-voltage lamps of various kinds. The low-energy and low-voltage light bulbs that have replaced them are more expensive to buy (though prices have and will come down), but last very much longer and use a lot less electricity.

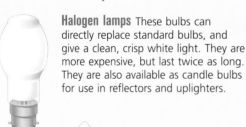

Halogen lamps These bulbs can directly replace standard bulbs, and give a clean, crisp white light. They are more expensive, but last twice as long. They are also available as candle bulbs for use in reflectors and uplighters.

Shaped bulbs

Used in wall and ceiling fittings, and available in a wide range of shapes, these lamps may have small bayonet caps (SBC) or small Edison screws (SES).

LOW-ENERGY BULBS

Compact fluorescent lamps (CFLs – page 96) and light-emitting diodes (LEDs – page 95) are designed to replace conventional light bulbs. They fit into standard bayonet or screw-in fittings, can use just a fifth of the energy of a filament bulb and last eight to ten times as long. Although more expensive to buy initially than conventional bulbs, they will save you a lot of money over time.

Striplights Tungsten filament striplights are ideal for use under shelves, inside cupboards and above mirrors. To avoid glare they can be used with a baffle.

Replacing a one-way switch or fitting a new dimmer switch

If a switch faceplate is cracked, you must replace it at once to prevent the risk of users touching live parts.

Before you start If you want to fit a metal switch, the switch cable must have an earth core connected to an earth terminal in the switch mounting box.
• Most light switches control just one light, and are one-way switches with two terminals on the back. Switches with three terminals are needed only for two-way switching arrangements (see right).
• You may want to replace an existing switch with a dimmer switch. Choose one that will fit the depth of the existing mounting box, and is suitable for the light(s) it will control.

Tools *Insulated screwdrivers (one with a small, fine tip).*
Materials *Light switch or dimmer switch.*

1 Turn off the main switch at the consumer unit and remove the fuse or switch off the MCB for the circuit you are working on.

2 Remove the screws securing the switch faceplate and ease it away from the mounting box. Keep the screws; the ones provided with the new switch will be metric and will not fit the lugs in an old pre-metric mounting box.

3 Use the fine-tipped screwdriver to release the switch cable cores from their terminals (see step 2, overleaf). Connect the red and black cable cores to the new faceplate. If there is an earth core in the cable, it will be connected to an earth terminal in the mounting box.

4 If you are fitting a metal faceplate to a metal mounting box, add a flying earth link (page 52) between the earth terminal on the faceplate and the one in the box. On a plastic mounting box, disconnect the switch cable earth core from the terminal in the mounting box and connect it to the faceplate earth terminals.

5 Fold the switch cable back into the mounting box and screw the faceplate to it, making sure that TOP is uppermost.

6 At the consumer unit, replace the circuit fuse or switch on the MCB and restore the power.

Replacing a one-way with a two-way switch

Two-way switches allow you to turn a light on and off from two switch positions. This is necessary for stairs and rooms with two doorways.

Before you start A length of three-core-and-earth cable has to be run between the two switches. A two-way switch has three terminals, usually marked L1 and L2 (or just 1 and 2) and COM (or COMMON). All two-gang switches are two-way types, but you can use them on a one-way system (using the common and L2 terminals).
• In three-core-and-earth cable, the cores are colour-coded brown, black and grey, and the earth core is bare as in other cables. There are no regulations about which terminals the three coloured cores should go to as long as the connections are the same in both switches. Usually the brown core is connected to the common terminals of each switch, the black cores to L1 and the grey cores to the L2 terminals.
• It is essential to tag the black and grey cores with brown sleeving or brown PVC insulating tape to show they can be live.

Tools *Suitable tools for preparing the route (page 57); insulated screwdrivers (one with a fine tip); trimming knife; wire cutters and strippers; pliers; pipe and cable detector.*

Materials *Two two-way switches; one new mounting box; screws and wallplugs; grommet; 1mm² three-core-and-earth cable; green-and-yellow plastic sleeving; brown plastic sleeving or brown PVC insulating tape; small and large oval conduit; galvanised nails.*

● **Cable core colours have changed (see page 56)**

1 Turn off the main switch at the consumer unit and remove the fuse or switch off the MCB protecting the circuit you are working on.

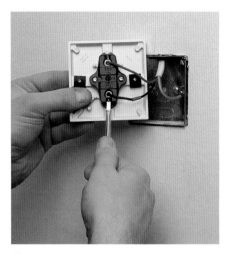

2 Undo the retaining screws that hold the existing one-way switch in place. Ease the faceplate away from the mounting box. Disconnect the two switch cable cores from their terminals and the earth core from its terminal in the mounting box. If the switch cable has no earth core, you must use only plastic switches and plastic mounting boxes.

3 Prepare the route (page 57) from the new switch position, up the wall, through the ceiling void and down the wall to the original switch position. In solid walls, cut a chase for the new cable and a recess for the new switch mounting box. Cut oval conduit to fit in the new chase, and feed one end of the three-core-and-earth cable into it. Run the cable through the ceiling void to a point above the existing switch.

4 Use a pipe/cable detector (page 11) to locate the position of the cable running down to the original switch. Enlarge the chase between switch and ceiling level without damaging the cable. Cut a length of large oval conduit to fit in the chase.

5 Pull the existing switch cable away from the mounting box and out of the newly widened chase. Feed it and the new three-core-and-earth cable into the conduit and on into the existing mounting box. Fix the conduit in the chase with galvanised nails.

6 Feed the three-core-and-earth cable into the new switch mounting box and fix the box in its recess.

7 Make good the wall around the new switch mounting box, and plaster over the two wall chases.

8 When the plaster has dried, prepare the cable for connection (page 39); remember to sleeve the bare earth cores. Use pliers to straighten the original cores.

9 At the original switch position, connect the red core from the original cable and the new black core to the terminal marked L1 on the new switch faceplate. Connect the red-tagged black core from the original cable and the new grey core to the terminal marked L2. Connect the new

brown core to the terminal marked COM. Screw the two green-and-yellow-sleeved earth cores into the terminal at the back of the mounting box.

10 If the new faceplate is metal, add a flying earth link (page 52) between the earth terminal in the mounting box and the earth terminal on the switch faceplate.

11 Fold the cables neatly into the mounting box and screw on the faceplate.

12 At the new switch position, connect the brown core to the terminal marked COM, the black core to L1, the grey core to L2, and the green-and-yellow-sleeved earth core to the terminal in the mounting box. Again add a flying earth link (page 52) if the new switch is metal.

13 Screw the new switch faceplate to its mounting box.

14 At the consumer unit, replace the circuit fuse or switch on the MCB, then restore the power and test the switches.

PLASTERBOARD WALLS

Running the wiring for two-way switching in hollow stud partition walls is much easier than for solid walls – as is cutting the hole to take the new flush drylining mounting box. See pages 60 and 81 for more details. Flying earth links are not needed, but metal switch plates should be earthed.

Moving a light switch

When you re-hang a door on the other side of a doorframe, the light switch will now be on the 'wrong' side. It's not too difficult to move it to the other side of the door.

Before you start Lift the floorboards in the room above the existing light switch to identify the cable that runs from the ceiling rose (or wiring junction box) and then down the wall to the switch. For an upstairs room, go into the loft – you may need to make a hole in the ceiling and poke a thin wooden stick through to identify the position.

Tools *Suitable tools for preparing the route (page 57); insulated screwdrivers (one with a fine tip); trimming knife; wire cutters and strippers; pliers.*

Materials *New plaster-depth mounting box; two short M4 screws; 5A four-terminal junction box; green-and-yellow earth sleeving; screws; red and brown sleeving (or insulating tape) oval conduit; galvanised nails; blanking plate (or wall filler).*

1 Turn off the main switch at the consumer unit. Remove the fuse or switch off the MCB of the lighting circuit you are working on.

2 Undo the retaining screws that hold the existing light switch in place. Ease the faceplate away from the mounting box, disconnect the cable cores from the switch and the earth core from the mounting box.

3 Go into the room above (or the loft) to the position where the switch cable comes through the ceiling. If the switch is mounted on a plasterboard partition wall, you should be able to pull the cable up through the ceiling. For solid walls, cut off the cable where it passes through the ceiling.

4 Prepare the free end of this cable (page 39) and connect the cores to three of the terminals of a 5A four-terminal junction box screwed to the side of a joist – make sure that the outer sheath of the cable finishes inside the junction box, tag the black core with red sleeving and cover the earth core with green-and-yellow sleeving.

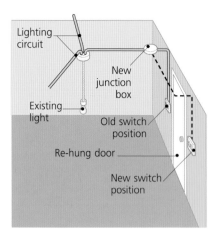

Lighting circuit

New junction box

Existing light

Old switch position

Re-hung door

New switch position

5 Prepare the route (page 57) from the new switch position, fit a new mounting box and run a length of 1mm² two-core-and-earth cable from the new switch position to the new junction box position in the same way as described for replacing a one-way switch on page 85.

6 Curl the cable inside the mounting box, make good the plaster round it and plaster over the wall chase.

7 At the junction box, prepare the end of the cable (page 39) and connect the brown core to the terminal with the existing red core, blue core (tagged with brown sleeving) to existing black and earth (covered with green-and-yellow sleeving) to the matching earth. The fourth terminal is left empty.

8 When the plaster repairs have dried, prepare the end of the cable at the new switch position, connect the brown core to the terminal marked COM and the blue core (tagged with brown sleeving) to the terminal marked L2 and the green-and-yellow sleeved earth to the terminal in the mounting box.

9 Fold the switch cable back into the mounting box and screw the faceplate to it, using new metric screws.

10 At the old switch position, cut off the cable. Unless you are prepared to fill the whole recess with wall filler (removing the old box if it protrudes) and redecorate, simply fit a blanking plate, using the old securing screws. With partition walls, you could remove the box and repair the plasterboard around that area.

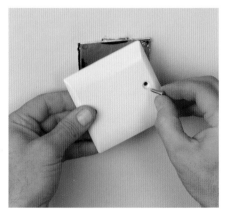

11 At the consumer unit, replace the circuit fuse or switch on the MCB, then restore the power and test the switch.

Fitting or replacing a ceiling switch

A ceiling-mounted pull-cord switch is the only type you are normally allowed in bathrooms, but is often fitted in bedrooms for two-way switching.

You will have to notify your local Building Control Department if you want to fit a new ceiling-mounted pull-cord switch in a bathroom (perhaps to replace a wall switch), but you do not need approval to install one in a bedroom.

For a lighting circuit, you need a 6amp pull-cord switch, available in both one-way and two-way versions. You will need a two-way version for two-way switching in a bedroom, but you could use it in place of a one-way switch if you needed to. The switch comes with its own plastic mounting box, which is secured to the ceiling.

Tools *Suitable tools for preparing the route (page 282); insulated screwdrivers (one with a fine tip); trimming knife; wire cutters and strippers; pliers.*

Materials *Ceiling-mounted light switch; 1mm² two-(or three-)core-and-earth cable; green-and-yellow earth sleeving; screws; spare piece of timber; red (or brown) sleeving.*

1 Turn off the main switch at the consumer unit and remove the fuse or switch off the MCB protecting the circuit you are working on – in this case, one of the lighting circuits.

2 Choose the position for the new pull-cord switch – if you are replacing a wall switch, make this more-or-less immediately above the existing wall switch. Make a hole in the ceiling large enough to pass cable through.

3 If you are replacing a wall switch with a ceiling switch, follow Steps 2, 3 and 4 for Moving a light switch (page 87), then go to Step 6 below. It is possible you may not need the junction box in Step 4 if there is sufficient cable to take through the ceiling to the pull-cord switch.

4 If you are replacing an existing ceiling switch, remove its cover, disconnect the wires and remove the base from the ceiling. Then go to Step 6 below.

5 If you are fitting a new two-way ceiling-mounted switch in a bedroom, follow the instructions on page 85, but take the new cable from the existing switch to the position of the new ceiling-mounted switch.

6 Make sure that there is adequate support for the ceiling light switch, using a timber batten as described on page 62.

7 Use a screwdriver to remove the knock-out in the base of the mounting box to make an entry for the cable.

8 Secure the mounting block for the switch to the ceiling, making sure that the screws go right through the ceiling into the timber block (and joist if appropriate).

9 Feed the end of the switch cable down through the hole and prepare the cores if necessary. Slip a length of brown sleeving over the blue wire.

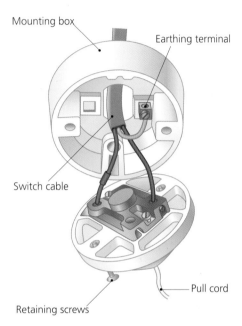

Mounting box

Earthing terminal

Switch cable

Pull cord

Retaining screws

10 Connect the cores to the terminals in the switch – brown to COM and blue to L1 or L2 for one-way switching, brown to COM, black to L1 and grey to L2 for two-way switching. If using the original cable (for straight replacement), it's red core to COM and red-tagged black to L1 or L2. Sleeve the earth core and connect it to the earth terminal in the mounting box.

11 Push the switch gently back into its mounting block and screw it in place. Do not overtighten the screws.

12 At the consumer unit, replace the circuit fuse or switch on the MCB, then restore the power and test the switch.

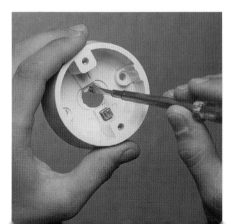

> ### HELPFUL TIP
>
> If the old ceiling switch has been painted over, run a trimming knife around the join between the ceiling and the switch before you remove it.

● Cable core colours have changed (see page 56)

Installing a new ceiling light fitting

Replacing a ceiling rose and pendant lampholder with a new light fitting gives your room an instant update.

Tools *Screwdriver; electrician's insulated screwdriver; bradawl.*

Materials *New light fitting; 5amp connector strip. Possibly green-and-yellow earth sleeving.*

1 Turn off the power to the lighting circuit at the consumer unit. Take down the bulb and lampshade, then undo the ceiling rose cover and unscrew the rose baseplate. Disconnect the single cable from the terminals – if there are two or three cables, see *Loop-in wiring* opposite. If the existing wiring has no earth core, you can only use double-insulated fittings.

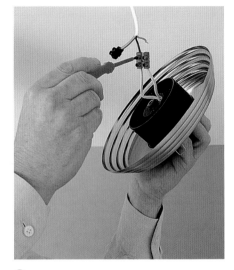

3 Remove the terminal connector block from the baseplate. Here, the fitting is double-insulated and needs no earth connection, so the circuit cable earth core should be terminated in a 5amp strip connector. Wire the circuit cable live and neutral cores into the terminal connector block.

2 After making sure that there is adequate support for the light (as for 'Fitting a ceiling rose' on page 62), hold the baseplate in position and mark the positions of fixing holes on the ceiling, using a bradawl.

4 Put the main terminal block and the strip connector containing the earth terminal back into the baseplate. Push the excess circuit cable up into the ceiling void.

5 While holding the baseplate and surround in their final position, locate the fixing screws in the holes you have already marked, and fix screws in place to secure the fitting to the ceiling. Add a light bulb and diffuser. Switch on the power supply.

LOOP-IN WIRING

Two or three cables present at the existing ceiling rose indicates loop-in wiring, when the switch cable is wired into the ceiling rose. If this is the case, identify and mark the cables (using the photograph on page 48 as a guide). If there is sufficient space, use four strip connectors wired up as above: (shown for clarity as the last light on a loop-in system with just one circuit cable): the circuit neutral core(s) and the switch return core linked to the light's flex tails; the circuit lives in the third terminal and all the sleeved earths in the fourth. If there is not space in the light fitting, you will have to re-connect all the cable wires in a junction box above the ceiling – as shown on page 92 (ignore the spur cable), taking a single cable back through the ceiling to the light.

Adding an extra light

You can run a spur cable from a lighting circuit to supply an extra lighting point.

There are three places where the spur can start: at any ceiling rose on a loop-in system (page 102), but easiest at the last one; at any box on a junction-box system (page 102); or at a new junction box inserted into the lighting circuit, whether it is a loop-in or a junction-box system.

Tools *Insulated screwdriver, tools for preparing the route (page 57); wire cutters and strippers; trimming knife; pliers.*

Materials *1mm² two-core-and-earth cable; green-and-yellow plastic sleeving; cable clips. Perhaps a three-terminal junction box to connect the spur cable.*

Preparation Turn off the main switch at the consumer unit and remove the fuse or switch off the MCB protecting the circuit you are working on. Prepare the route (page 57) and lay the new cable.

Adding to the last loop-in ceiling rose on the circuit

1 Unscrew the loop-in ceiling-rose cover and slide it down the flex. Then loosen the screws holding the rose baseplate to the ceiling. This will make it possible to work the new spur cable carefully through the ceiling in the same hole as the existing circuit and switch cables and through the entry hole in the ceiling rose.

2 Prepare the new cable for connection (page 39). Screw the rose baseplate back to the ceiling.

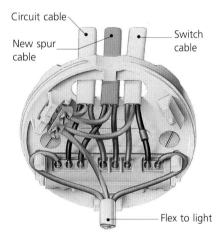

Circuit cable

New spur cable

Switch cable

Flex to light

3 Connect the brown core of the spur cable to the spare terminal in the central terminal block, which already holds two red cores. Next, connect the blue core to the spare terminal in the outer terminal block to which the existing circuit cable black core is connected. Finally, connect the green-and-yellow-sleeved earth core to the separate earth terminal.

4 Screw the ceiling rose cover back in place.

● Cable core colours have changed (see page 56)

Adding a spur to a junction box wiring system

1 Remove the cover of the junction box where the spur will start. You can cut out an entry hole in it for the spur, or let the spur share the same entry hole as an existing cable.

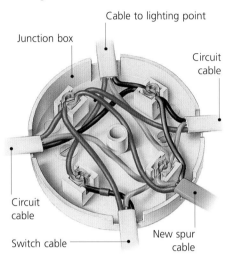

Junction box

Cable to lighting point

Circuit cable

Circuit cable

Switch cable

New spur cable

2 Prepare the new spur cable for connection (page 39). Feed it into the box.

3 Connect the brown core of the new spur cable to the terminal where two or three red cores are already connected. Connect the blue core to the terminal where two or three black cores are already connected. Connect the green-and-yellow-sleeved earth core to the terminal where the other earth conductors are already connected.

4 Screw the cover on the junction box.

Leading a spur from a new junction box in the circuit

1 Trace the cable to which you plan to connect the spur, to make sure that it is not a switch cable but the circuit cable running from one loop-in rose or junction box to the next.

2 Screw the baseplate of the new junction box to the side of a joist so that the circuit cable runs across it.

3 Cut the cable over the centre of the box, then strip back enough of the outer sheath to allow the cores to reach their terminals with the sheath still within the box. Then strip about 15mm of the core insulation from the live and neutral cores.

4 Sleeve the bare earth cores with green-and-yellow sleeving.

5 Prepare the ends of the spur cable for connection (see page 39).

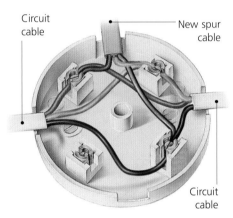

Circuit cable

New spur cable

Circuit cable

6 Connect the circuit and the new spur cable cores to the terminals as shown here. Then screw the cover back onto the junction box.

At the new lighting point

At the new loop-in ceiling rose, connect in the spur cable cores – the brown core to the central terminal, the blue core to one of the outer terminals and the sleeved earth core to the earth terminal. Connect the flex cores to the outer terminals. Run the switch cable to the new switch position and connect it to the switch terminals.

Next, connect the switch cable at the rose. The brown core goes to the centre terminal, the blue core to the outer terminal where the brown flex core is connected and the earth core to the earth terminal. Identify the blue core as live by wrapping some brown PVC tape around it.

Adding a spur for a wall light

You can add a wall light by leading a spur from an existing lighting circuit.

Before you start The wall light may incorporate its own switch; if it does not, or if you want to be able to switch the wall light from the door, you can fit an extra junction box in the spur cable to connect a new switch cable.
• You should fit a mounting box in the wall for making the wiring connections.
• If the base of the light fitting is too small to conceal a square box, use a round, plastic or metal conduit box or a slim architrave mounting box. Use a round dry-lining box in a timber-framed partition wall.
• Some wall lights have baseplates with screw fixing holes to match the holes in circular conduit or drylining boxes.
• Choose a light wired with an earth conductor or double-insulated (marked ▣).

Wiring to a wall light

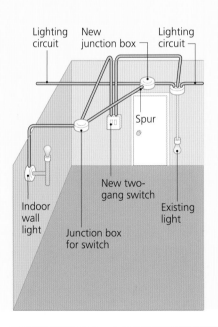

Lighting circuit

New junction box

Lighting circuit

Spur

New two-gang switch

Indoor wall light

Junction box for switch

Existing light

Tools *Suitable tools for preparing the route (page 57); insulated screwdrivers; trimming knife; wire cutters and strippers; pliers. Perhaps a circuit tester.*

Materials *Wall light; 1mm² two-core-and-earth cable; green-and-yellow earth sleeving; metal mounting box, conduit box or slim architrave box or round dry-lining box; short length of green-and-yellow-insulated earth conductor; connector block with three pairs of terminals (cut from a strip). Junction boxes for connecting the spur and switch; brown plastic sleeving; two-gang light switch.*

Preparation Turn off the power at the main switch in the consumer unit, and remove the fuse or switch off the MCB protecting the circuit you will be working on. Prepare the route for the spur cable. You will need to lead it above the ceiling from a convenient spot on the lighting circuit to a point immediately above the wall light position, and then down the wall.

The wall light should be about 1.8m above floor level. Make a recess there for the mounting box or slim architrave box, or cut a round hole in a plasterboard wall for a dry-lining box.

In order to incorporate a switch, lead the spur cable route above the ceiling to a spot halfway between the wall light position and the switch position and fit the four-way junction box there. Continue the cable run from there to the wall light position and run another cable from the junction box to a point directly above the switch. You should be able to complete the route by enlarging the recess in the plaster leading down the wall to the existing switch. Take great care not to damage the existing switch cable. Unscrew and disconnect the switch. You can use the single mounting box already fitted and fit a two-gang switch over it when you are ready to make the connections.

Alternatively fit a new switch next to the existing one.

1 Lay the spur cable along the route from the circuit to the junction box for the switch, and from there cable to the wall light position and also to the switch position. Secure the cable along the route.

2 At the wall light, feed the cable into its mounting box. At the switch, feed the switch cable into its mounting box.

3 Prepare the new cable ends for connection. Slide a piece of brown sleeving on the switch cable's blue core at each end.

4 Connect the spur to the lighting circuit (page 91).

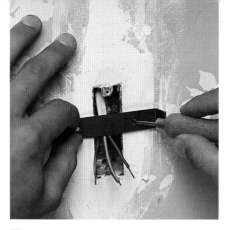

5 Hold the wall light or its bracket in place and push the bradawl through its fixing holes to see if they match up with the screwed holes of the mounting box. If they do not align – or if you are using an architrave box – make drilling marks with the bradawl. Drill at the marks and plug the holes.

6 Make good any plaster damage.

Connections at the wall light

1 Connect the brown and blue cable cores to the two outer terminals on one side of the terminal connector block.

2 At the central terminal, connect the cable earth core. If you are using a metal mounting box, connect a flying earth link (page 52) to the central terminal, on the same side. Connect the other end of this earth link to the earth terminal in the box.

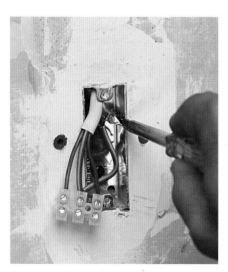

3 Connect the flex cores from the wall light to the terminal connector block with brown, blue and green-and-yellow cores opposite the brown, blue, and green-and-yellow cable cores respectively.

4 Screw the wall light into place to or over the mounting box.

Connections at the switch

1 Make the connections behind the new two-gang switch. All two-gang switches are designed to operate as two-way switches. When you use one as part of a one-way system, you must use the Common terminal and one of the other two terminals. Connect the new brown and brown-sleeved black cores to the first switch gang using the terminals marked Common and L2. Connect the cores from the original cable to the second switch gang in the same way.
  Screw both the green-and-yellow-sleeved cores into the earth terminal at the back of the box.

2 Fold the cables neatly and screw the switch faceplate to the mounting box.

3 At the junction box for the switch, connect the brown cores from the spur and switch cables to one terminal.

4 Connect the blue cores from the spur cable and the cable leading to the wall light to another terminal.

5 Connect the brown core from the wall light cable and the brown-sleeved black core from the switch cable to a third terminal.

6 Finally, connect all the green-and-yellow-sleeved earth cores to the fourth terminal.

7 At the consumer unit, replace the circuit fuse or switch on the MCB and restore the power.

Choosing low voltage lighting

Low voltage lighting uses a transformer to step down mains electricity to just 12 volts. A transformer can be attached to each individual light, or one transformer can supply several lights. A transformer can be large and heavy, so mounting it away from the lights often makes sense. One major advantage of low voltage lighting is that the risk of electric shock is minimal – it is quite safe to touch the terminals on the low voltage side of the transformer.

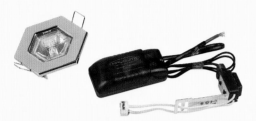

Separate transformers Having a separate transformer makes the light fitting compact and easily fitted into a confined space. A single transformer can run several lights, or each light may have its own box. The transformer can be mounted near the mains power supply, allowing unobtrusive cables to run to the light fitting or it can be slotted into the ceiling void alongside the light. Compact and maintenance free, transformers are a fit-and-forget item that can be hidden away on top of a kitchen cupboard or in a ceiling or wall space.

Integral transformers The outward appearance of these lights is similar to those that have a separate transformer, but there is a transformer built in to each light fitting. They are generally more expensive than those with a remote transformer as not only are you paying for the cost of the fitting but also the cost of a transformer for each light source. Also, while each light fitting may be small, its transformer may be fairly bulky and this must be taken into account when siting lights.

Dimming If low voltage lights are to be dimmed then the dimming control must be mounted on the primary or 230 volt side of the circuit (not the transformer side). Dimming increases lamp life for low-voltage lighting, as it does for regular light bulbs.

Colour All light has what is known as a colour temperature. The higher the number, the whiter the light appears. Many low voltage lights used in shops for display purposes use halogen bulbs, which appear too white and stark for domestic use. For room lighting, low voltage bulbs give a warmer more relaxing light.

ENERGY SAVING LIGHTING

Although more expensive at the outset, low voltage lights use only around a third of the electricity used by conventional lights and last much longer – typically 2-3,000 hours.

CHOOSING LED LIGHTING

Light emitting diodes (LEDs) are familiar enough in electronic equipment, but are now being produced in a wide range suitable for use in the home. Although very expensive to buy, they use very little electricity, don't emit heat and last for a very long time (up to ten years). LEDs are available to replace most types of light bulb and come in a range of exciting colours as well as white. Most can be used in standard light fittings.

You can also fit walk-over LEDs into floors, both indoors and out. Because these lights remain cool to the touch, they are safe for use in playrooms.

Choosing fluorescent lamps

A fluorescent lamp uses electronic discharge to illuminate the coating of a tube. It gives roughly five times as much light per watt as a tungsten light bulb and lasts at least ten times as long.

Fluorescent tubes Long white tubes, for use in fluorescent light fittings with a starter, are found in many homes. However, they now come in many sizes and more colours than the traditional clinical white. Common sizes are 26mm diameter (labelled T8) and 16mm (T5). Less efficient 38mm (T12) tubes are still sold for older fittings. Power output depends on length: 600mm is 18W, 1200mm is 36W and 1500mm is

58W, for example. The most efficient type is the triphosphor fluorescent tube, which also has improved colour rendering and a life of up to 20,000 hours. Cannot normally be used with a dimmer switch.

Compact fluorescent lamps Also known as low-energy and energy-saving lamps. These work in the same way as fluorescent lights, but the tube is bent back on itself and the control gear is built in, so they can be used to replace normal light bulbs with a screw or bayonet fitting.
• An 11W CFL replaces a 60W light bulb
• A 20W CFL replaces a 100W light bulb
• Many types are dimmable

Other options Fluorescent lamps come in a range of shapes, including pencil shapes, bent tube shapes and glass covered lamps, which look similar to the light bulbs they are meant to replace. Low-energy replacements are also available for R50, R63 and R80 reflector bulbs – see page 84. Flat square (2-D) lamps have their own range of special fittings including porch lights, bulkhead lights and ceiling lights.

Fitting plug-in under-cupboard lighting

The simplest way to illuminate your worktops is with a chain of linked fluorescent strip lights that are plugged in to an electric socket.

This job does not involve mains wiring, so can be done without giving notification. Once the lights are in place, plug the master light (the first in the run) into a mains socket to power the lights. Each light has a rocker switch. Make sure the plug has a 3amp (red) fuse, not a 13amp (brown) one.

The striplights are sold in kit form – with fixing clips, screws, cable clips and two-core connecting flexes that simply plug in to link the run of lights together.

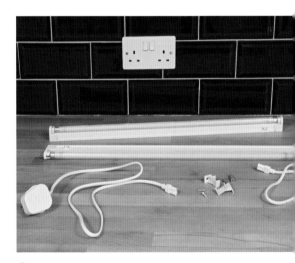

1 Allow one fitting for every 500mm of worktop. Lay the kit out on the worktop to plan where to fit the lights. Ensure there is a socket outlet close to the master light.

2 Position the fixing clips and screw them to the underside of the wall units.

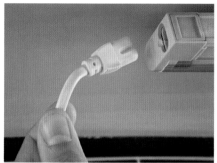

3 Plug in the connecting cables to link all the lights together.

4 Secure the lights into position using their fixing clips. Pin excess flex to the underside of the cupboards using plastic cable clips.

5 Plug in and switch on at the mains socket, and test each light in turn.

Wiring in under-cupboard kitchen lights

It is possible to connect under-cupboard lights permanently to the kitchen ring circuit. You can use fluorescent or tungsten striplights or special low-voltage spotlights.

Low-voltage spotlights under kitchen cupboards need a separate transformer. This can run more than one light – a 60 volt-amp transformer, for example, can power three 20W lights. The transformer can be wired to a switched fused connection unit (FCU) wired into the kitchen ring circuit. The switch turns the lights on and off.

Tools *Suitable tools for preparing the route (page 57); insulated screwdrivers (one with a fine tip); trimming knife; wire cutters and strippers; pliers.*

Materials *Switched fused connection unit; flex outlet plate; mounting boxes; 2.5mm² and 1.5mm² two-core-and-earth cable; 0.75mm² flex; green-and-yellow plastic sleeving; screws; low-voltage light fittings; 12V transformer.*

1 Turn off the main switch at the consumer unit and remove the fuse or switch off the MCB protecting the kitchen ring circuit.

2 The FCU can be wired into the kitchen ring circuit (page 66) or wired as a spur from a kitchen socket outlet (page 63).

3 Prepare the route for the 2.5mm² cable running to the position of the switched FCU. Fix the switch mounting box in place and run the cable from the socket outlet.

4 A good place for the transformer is on top of the kitchen cupboard. Prepare a route for the 1.5mm² cable from the FCU to a position above the cupboard and fit a mounting box for the flex outlet plate (page 53). Repair all wall damage.

5 To run the wires from a light fitting to the transformer, drill a hole in the base and top of the cupboard at the light position and feed a length of stiff wire or a flexible rod up through the two holes to the top of the cupboard. Pass the fitting over the end of the stiff wire (or rod) and tape the cable of the light to the end of the stiff wire (or rod) and pull it through to the transformer. Repeat for all the fittings.

6 Screw each light fitting to the underside of the cupboard and position the lampholder in the light fitting.

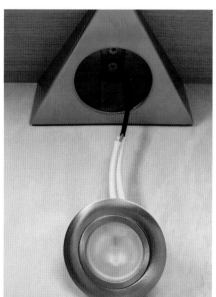

7 At the transformer, connect the cables from the lights to the output terminals and connect the 0.75mm² flex to the input terminals, following the instructions that have been provided.

8 Make the connections at the socket outlet for the spur cable to the FCU (page 54), at the FCU for the spur cable and cable leading to the flex outlet plate (page 65) and at the flex outlet plate for the cable from the FCU and the flex leading to the transformer (page 68).

9 Secure the faceplates to their mounting boxes, and fold the cables neatly into them.

10 Turn the electricity back on at the consumer unit and test the lights.

Wiring in fluorescent lights

It is not always feasible or desirable to fit a lighting kit that runs from a mains socket or an FCU. You may need to wire a fluorescent lamp fitting into the mains lighting circuit. If you are in any doubt about tackling this work, then hire a qualified electrician.

Fluorescent lights make good task lighting: a lamp above a desk or a dressing table, for example. They can also be used as accent lights, to illuminate a picture or another feature in a room. In all these instances, you will want to hide the wiring. The light itself can be operated with a pull-cord or rocker switch, or a wall-mounted light switch, usually sited near the door.

Tools *Bradawl; tools for preparing the route (page 57); trimming knife; wire cutters and strippers; insulated screwdrivers; pliers.*

Materials *Light fitting; two-core-and-earth 1mm² cable; green-and-yellow earth sleeving; cable clips. Perhaps also four-way junction box; one-way light switch with mounting box, fixing screws; grommet; brown plastic sleeving; insulating tape.*

Before you start To connect the fitting to the electricity supply, you will need to lead a spur to it from the lighting circuit (page 91). Remove any baffle and end covers from the fitting. Hold the base plate against the wall (or shelf) and mark the fixing points and, if necessary, the cable entry hole. Drill holes at the marks and screw the base plate of the light fitting firmly in place.

Fitting a light with a built-in switch

1 Turn off the main switch at the consumer unit and remove the fuse or switch off the MCB protecting the circuit you are working on.

2 Lead a spur from the lighting circuit (page 91) to the position of the light fitting, burying it in the wall plaster or hiding it behind plasterboard.

3 When plaster repairs have dried, thread the 1mm² spur cable through the entry hole of the light fitting. Prepare the cable ends (page 39) and connect the cable cores to the terminals of the terminal connector block in the light fitting. Connect the brown core opposite the brown flex core, the blue core opposite the blue flex core, and the green-and-yellow sleeved core to the terminal marked E or ⏚. If the fitting is double-insulated (marked ▣) and has no earth terminal, cut the earth conductor of the cable and seal it with insulating tape or fit a single plastic terminal block (cut from a strip) to the end of the sleeved earth core.

4 Screw the end covers in place and fit the baffle or shade.

5 Turn the electricity back on at the consumer unit and test the light fitting.

Fitting a light with a separate switch

1 Turn off the main switch at the consumer unit and remove the fuse or switch off the MCB protecting the circuit you are working on.

2 Lead a spur from the lighting circuit (page 91) to a junction box in the ceiling space or loft above the position for the light.

3 Prepare the cable routes. You will need one cable from the junction box to the light fitting and another from the junction box to the switch position. Fit a mounting box.

4 Lay 1mm² cable from the junction box to the light. Feed it into the light fitting through the entry hole.

5 Then lay 1mm² cable from the junction box to the light switch position and feed it into the mounting box. Repair plaster.

6 Prepare the cable ends for connection. Remember to put a piece of brown plastic sleeving over the blue core of the switch cable at both ends to show that it is live.

7 Connect the cable to the terminal block in the light fitting (left, step 3).

8 At the switch, connect the brown core to one of the terminals and the brown-sleeved blue core to the other terminal. Connect the green-and-yellow sleeved earth core to the earth terminal in the mounting box. Screw the switch faceplate to its mounting box, folding the cable neatly into the box.

9 At one terminal in the junction box, connect the brown cores of the spur cable and the switch cable. At another terminal connect the blue cores of the spur cable and the cable from the light. At a third terminal connect the brown core of the cable from the light and the brown-sleeved blue core of the switch cable. At the fourth terminal, connect all the green-and-yellow sleeved earth cores.

Fitting low-voltage recessed spotlights

Brilliant, low-voltage lights are ideal for kitchens. Because they are small, it is possible to site lights directly above where they are needed, thus avoiding the shadows that can arise with centrally mounted lighting.

The following instructions relate to a typical installation of ceiling-mounted recessed lights to replace an existing single light. Not all lights are identical, so be sure to read the installation instructions that come with your lights, as there may be differences from those described here. In order to fit these lights, you will need access to the ceiling from above.

Tools *Pencil; padsaw, or power drill with holesaw attachment, probably 57mm in diameter; compasses; insulated screwdriver; trimming knife; wire cutters and strippers.*

Materials *Low-voltage lights with individual transformers; 1mm² two-core-and-earth cable; junction boxes.*

Before you start Although low-voltage lighting is very safe, the transformers connect to the 230-volt mains circuit and must be treated accordingly.

1 Decide where you want to site the lights and draw circles on the ceiling the same diameter as the light fittings. Then use a holesaw of this diameter – usually 57mm – fitted to a power drill, to cut perfect holes.

Alternatively Drill a hole just inside each circle large enough to admit a padsaw blade and cut the holes by hand.

2 Switch off the power at the mains and remove the fuse or switch off the MCB protecting the circuit. Examine the wiring to the present light fitting. If there is just one cable leading into a ceiling rose or the light fitting, disconnect this and reconnect the cores into three of the four terminals of a

MAINTENANCE AND REPAIRS

5A junction box secured to the side of a joist. If you have a loop-in ceiling rose, identify the cables (page 47), label them with a pen and re-connect them in a 15A junction box, secured to a joist, using all four terminals (left hand picture on page 92 ignoring the spur cable).

3 Remove the old light fitting and make good the damage to the ceiling plaster.

4 Connect a new length of 1mm² two-core-and-earth cable to the terminals in the junction box – blue core to black circuit cables and brown core to red sleeved black core in switch cable (or, for a single cable, blue to black and brown to red). Run the new cable to the nearest light position. Terminate it at a junction box together with a new length of 1mm² two-core-and-earth cable. Run the cable to the next light in the sequence and so on until each hole has wiring to it and a junction box to connect into. Working from above, use a cordless drill with a 12mm wood bit if you have to make cable holes through any joists.

5 At each location, wire in the transformer supply lead to the junction box.

HELPFUL TIP

Use a stud detector (page 11) to check the ceiling joists in the area that you wish to place the lights and mark these with a pencil. Ideally, lights should be placed centrally between the joists. If you position most of the lights between the same two joists, you will have less trouble running cables as you will not need to drill through joists for the cables.

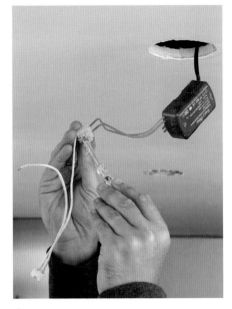

6 Connect each light to its transformer, using the plastic strip connectors supplied. Push each transformer and junction box into the ceiling void, leaving the bulb connectors hanging down.

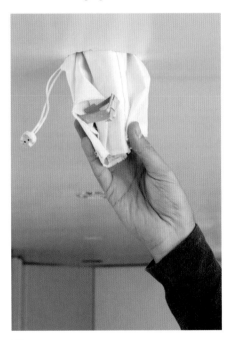

7 Push the smoke hoods up into the ceiling void, taking care not to push the bulb connectors with them.

8 Insert the light fittings into the holes until you hear them click: as you push the light up, the shorter arms on the clips force the longer arms to flatten against the upper surface of the ceiling.

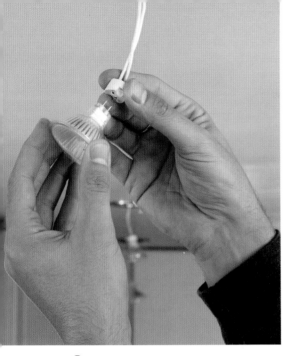

9 Fit a halogen bulb into each light fitting. Push the bulbs into place and snap on the metal ring clip to hold them. Turn on the power at the main switch to check the operation of the lights. If you need to change a bulb at a later date, gently squeeze the ends of the metal ring clip together so you can remove it and take out the bulb. Be sure to replace it with one of the same wattage.

SAFETY WARNING

Each recessed spotlight should be fitted within a smoke hood. This sits in the ceiling void above each spotlight and prevents smoke from the kitchen rising through gaps in the ceiling and floorboards to the upper floors, in the event of a fire.

✋ Wiring a new lighting circuit

Lighting circuits are wired with 1mm² two-core-and-earth cable. A house usually has two lighting circuits, one serving the downstairs and the other upstairs. You may need an extra circuit if you want to install a lot of extra lighting, or if you build an extension.

A lighting circuit is protected by a 5amp fuse or a 6amp MCB at the consumer unit. The maximum load that one circuit can supply is 1200 watts. With modern energy-saving bulbs, this limitation is no longer a problem.

In order to meet the latest Wiring Regulations, a new lighting circuit installed by you with cable run in walls will have to be protected by a 30mA RCD. Unless you have a spare 6A MCB that is already RCD-protected (or can be replaced with a 6A RCBO – see page 72), install a 30mA RCD and a 6A MCB in a new four-way enclosure as described on page 55.

For each lighting point on the circuit, the cable passes through either a loop-in ceiling rose or a junction box and is connected to terminals there. From the rose or junction box, wiring leads off to the light and to the light switch. In many houses, a light circuit will include both loop-in ceiling roses and junction boxes.

Loop-in system

A loop-in ceiling rose has a row of terminals to connect in three cables – the two circuit cables, and the switch cable – and the flex that goes to the lampholder. If the rose is the last one on the circuit, there will be only one main circuit cable.

All the connections, except the earths, are made to a single row of terminals. All can be reached simply by unscrewing the rose cover. There is no need to lift floorboards in the room above to reach them, as with junction boxes.

Junction box system

The circuit cable is not led to the lighting point but to a junction box between the lighting point and the switch position. The junction box has four terminals for connecting four cables – two circuit cables, the switch cable and the cable to the lighting point.

• This system is seldom used for a whole lighting circuit because of the extra work involved in installing a junction box for each lighting point, and the fact that once the wiring is completed the connections are not accessible without lifting a floorboard. But it is needed for individual lighting points such as strip lights, spotlight tracks, recessed lights and other special fittings which are not designed for loop-in wiring; instead these incorporate a terminal connection block for connection to a single cable from a junction box.

Planning the circuit

Consider whether each lighting point will be better suited to loop-in or junction box wiring. Position any junction boxes in a ceiling space or loft where you can screw them to joists.

• Cables to light switches must run from the rose or junction box to a spot directly above the switch and then be chased in plaster, or led inside a stud partition wall, or in trunking vertically down the wall. Never run cables diagonally across walls; there is a danger of piercing them later with fixings for pictures or built-in furniture.

Installing the circuit

Tools *Suitable tools for preparing the route (page 57); trimming knife; wire cutters and strippers; insulated screwdrivers; pliers.*

Materials *$1mm^2$ two-core-and-earth cable; green-and-yellow sleeving; brown sleeving; ceiling roses; flex; lampholders; light switches with mounting boxes, screws and grommets; available 5amp fuseway or 6amp MCB in consumer unit. Perhaps $1mm^2$ three-core-and-earth cable for two-way switches; special light fittings.*

1 Prepare the route for the circuit cable (page 57). Leave one end at the consumer unit for connection later, and run the cable up to ceiling level. If it is to go above the ceiling, you will have to lift floorboards to run it along and through joists.

2 Prepare the routes for the switch cables which will run from every loop-in-ceiling rose or junction box to the switch positions. If a light is to be controlled by two-way switches (page 85), prepare a route to the second switch position.

3 For lighting points that will be wired on the junction box system, prepare routes from the junction boxes to the light positions.

4 Fit ceiling roses securely (page 62) where there are to be pendant lights.

5 Fit junction boxes where necessary for special light fittings, screwing the bases of each to the side of a joist.

Loop-in wiring system

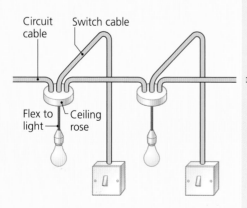

Circuit cable

Switch cable

Flex to light

Ceiling rose

Junction box wiring system

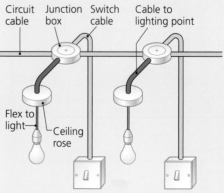

Circuit cable

Junction box

Switch cable

Cable to lighting point

Flex to light

Ceiling rose

6 Fit mounting boxes for all the switches.

7 Lay lengths of cable from above the consumer unit to the first ceiling rose or junction box, from there to the next rose or box, and so on. Leave enough cable at the ends of each length for the cores to reach all the terminals easily. Secure the cable by threading it through joists or by fixing clips to hold it to the sides of joists.

8 Lay and secure switch cables from each loop-in ceiling rose or junction box to each switch. For any two-way switches, also lay three-core-and-earth cable to connect the two switches. Allow enough cable to reach all the terminals easily.

9 Feed the cable ends into ceiling roses or junction boxes and into the light switches. Repair all plaster damage.

10 Prepare the cable ends for connection. Sleeve the earth conductors and on all switch cables put brown sleeving over the blue conductor at both ends.

11 For each ceiling rose, cut a length of flex that will allow the lampholder to hang at the desired height. Connect the flex to the lampholder first, then slide the cover of the ceiling rose onto the flex and connect it to the rose (see page 47).

Connecting at switches

Connect one-way and two-way switches by following the procedures described on page 85.

Loop-in connections at a ceiling rose

At each ceiling rose except the last one on the circuit, two circuit cables have to be connected; at the last there is only one circuit cable. There is also a switch cable and a light flex to connect at each rose.

1 In the central group of three terminals, connect the brown cores from the two circuit cables and switch cable. Connect the blue cores from the circuit cables to one of the outer terminals. Leave the outermost one clear for connecting the blue flex core.

2 At the other outer terminal, connect the brown-sleeved black switch core. Connect the brown and blue flex cores to the outer terminals and pass the flex cores over the hooks to take the lampshade's weight.

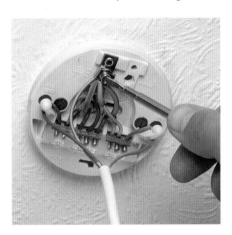

3 Finally, connect all the green-and-yellow-sleeved earth cores to the terminal marked E or ⊕. Fit the ceiling-rose cover.

Connections at a junction box

At each junction box except the last on the circuit there are two circuit cables, a switch cable and a cable to the lighting point. At the last box there is one circuit cable.

1 At one terminal, connect the brown cores from the two circuit cables and the switch cable. At another terminal, connect the blue cores from the two circuit cables and the cable to the lighting point.

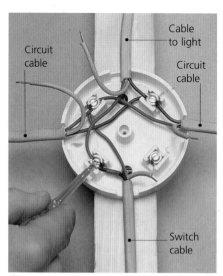

Circuit cable

Cable to light

Circuit cable

Switch cable

2 At a third terminal, connect the brown core from the cable to the lighting point and the brown-sleeved blue switch core.

3 At the fourth terminal, connect all the sleeved earth cores.

4 Screw on the junction box cover.

Connecting light fittings to a junction box system

Single spotlights, multi-spotlight tracks, recessed lights and fluorescent strip lights are among the special light fittings that are best connected to a junction box system.

1 Fix the light fitting to the ceiling. The mounting plates for tracks and strips should be screwed through the plasterboard or plaster into joists or into battens nailed between the joists (see page 62).

2 If the light fitting has a terminal block holding the flex cores and ready to receive the cores from the light cable, connect the brown cable core to the terminal opposite the brown flex core. Connect the blue core to the terminal opposite the blue flex core. Connect the green-and-yellow-sleeved earth core to the third terminal. If the light fitting is double-insulated (marked ▣) there will be no earth terminal; cut short the cable earth core and seal it with insulating tape so that it cannot touch any terminals.

Alternatively The light fitting may have no terminal connector block – and no space to fit one. Instead, install a circular drylining box in the ceiling and make the cable and flex connections inside that, using a terminal connector block cut from a strip.

Connecting at the consumer unit

1 Turn off the main switch on the consumer unit, if you have not already.

WARNING

The main switch disconnects the fuses or MCBs and the cables leading from the consumer unit to the household circuits. It does NOT disconnect the cables entering via the meter from the service cable. Do not interfere with these – they are live at mains voltage.

2 Remove the screws from the consumer unit cover and open it. Feed the circuit cable into the top of the consumer unit. Use an existing hole if there is room.

3 Check that you have stripped off enough outer sheath for the cores to reach the terminals easily, but make sure that the sheath enters the consumer unit.

4 Connect the new cable's brown core to the terminal at the top of the spare 6amp MCB or RCBO you are using.

Alternatively Connect the brown core to the top of the 6A MCB in the four-way enclosure as shown on page 55. The other MCB in this enclosure could be used for a new cooker circuit (page 72), a new shower circuit (page 69) or a new socket outlet circuit (page 76).

5 Connect the new cable's blue core to the neutral terminal block. Then connect the green-and-yellow-sleeved earth core to the earth terminal block. If you are using a spare fuseway, fit a 5amp fuse. Switch on MCB or RCBO.

Alternatively Connect the blue core to the neutral terminal block in the four-way enclosure and the green-and-yellow sleeved earth core to the earth block.

6 Fix the cover of the consumer unit back in place, refit the cover over the fuse carriers and turn the main switch back on.

Testing and making good

Fit bulbs or fluorescent lamps into the lampholders and fittings and turn on each light in turn to make sure that it is working. If not, check and tighten the connections, remembering to turn off the electricity at the consumer unit before you do so.

Adding an aerial socket outlet for a second TV set

You can connect more than one TV set to your aerial by leading the aerial cable into a splitter unit, which has one input and two outputs.

Run coaxial cable from each output on the splitter to a coaxial socket outlet near each TV set. Link each set to a socket with coaxial cable fitted with a coaxial plug at each end. There are male and female plugs to suit different socket outlets.

The aerial cable may be run outside the house and brought inside through a hole drilled in a wall or window frame. Position the splitter where the cable enters the house. The cables between splitter and sockets are normally run as surface wiring, typically along the tops of skirting boards, held in place with matching cable clips or hidden inside mini-trunking.

Where the coaxial cable runs into the house from outside, you can use flush-fitting socket outlets over standard single mounting boxes recessed into the wall; for surface wiring, use surface-mounted sockets and boxes screwed to the wall.

Tools *Suitable tools for preparing the route (page 57); trimming knife; wire cutters and strippers; insulated screwdrivers; pliers.*

Materials *Coaxial TV cable; cable clips or trunking; splitter unit; two flush-fitting coaxial socket outlets with mounting boxes and grommets, or two surface-mounting coaxial sockets; coaxial plugs.*

1 Prepare cable routes from the position for the splitter to the position for the sockets and, if necessary, from where the cable enters the house to the splitter.

2 If you are using flush-fitting socket outlets, fit mounting boxes for them. Knock out a convenient entry hole in each box, fit a grommet and screw the boxes in place. Feed the aerial cable out through the wall to the chosen position for the splitter.

3 Lay cable along each route from the splitter position to the socket outlet positions. Use cable clips or trunking to secure it. Avoid bending it sharply.

4 If you are fitting flush socket outlets, feed the cable end for each outlet through the grommet into the mounting box. Make good the plaster and wait for it to dry.

Alternatively For surface-mounted sockets, feed the cable end up through the floor at the base of the skirting. You will have to cut a notch or drill a hole in the floorboard to prevent it from chafing the cable. If the cable is not under the floor, feed it from the skirting or trunking to the spot where the base of the socket will be.

5 Prepare all the cables ends for connection. Make a 30mm slit with a trimming knife lengthwise at the end of the outer sheath. Fold back the sheath and cut it off. Loosen the wire mesh and press or fold it back to leave 20mm of inner insulation. Use wire strippers to remove 15mm of the inner insulation.

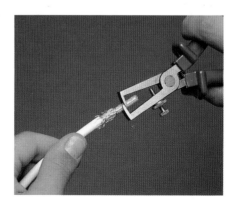

6 To connect at a flush socket outlet, secure the inner wire tightly in place at the terminal. Loosen the adjoining clamp, secure the wire mesh and cable under it and screw the clamp tight. The strands of the mesh must not touch the inner wire. Screw the socket outlet over the mounting box.

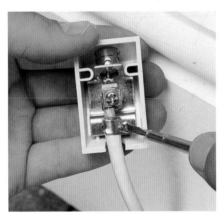

Alternatively With a surface-mounted socket outlet, secure the inner wire to the central terminal and the wire mesh and cable itself under the clamp, and screw the socket to the skirting board behind the TV set.

7 Secure the splitter to a joist or to the skirting and push the aerial cable plug into the correct socket on the splitter.

8 Plug the cables that run to the sockets into the splitter outlets, and the other two cables into the socket at one end and the TV set at the other end.

Choosing connectors for aerials

Socket for surface fixing on skirting

Splitter for over or under floor

Surface-mounted splitter

Flush socket to fit over mounting box

Female plug

Male plug

Coaxial cable

Wiring a coaxial plug

Aerial connectors can work loose over time, or you may need to extend the cable to reach your television. You can wire a new plug to a coaxial cable or repair a broken connection.

Tools *Trimming knife; pliers.*

Materials *Required length of coaxial cable; pair of coaxial plugs.*

1 Strip away a 30mm length of the outer sheath to expose the wire mesh beneath. Fold back 20mm of the wire mesh to expose the inner insulation.

2 Use wire strippers to remove 15mm of the inner insulation (see page 105), leaving the central inner wire exposed. Slide the cap down the cable and fit the cable grip over the exposed mesh and cable. The strands of the mesh must not touch the inner wire. You will have to open the jaws of the grip to cover the mesh and then squeeze them together with a pair of pliers.

3 Feed the inner wire into the pin moulding and fit the moulding in the plug body. Slide the cap over the cable grip and screw it to the plug body. (For a really secure connection, solder the inner wire into the pin moulding.)

4 Connect all the plugs in the same way, fitting a male or female plug to suit the connection on the socket outlet or splitter and the television set.

COMPONENTS OF A COAXIAL PLUG

Coaxial plugs come in several parts, which must be separated for fitting. The wire at the centre of the cable carries the signal and connects with the pin in the male plug.

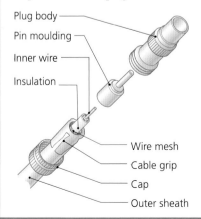

Plug body
Pin moulding
Inner wire
Insulation
Wire mesh
Cable grip
Cap
Outer sheath

Splitting and combining signals for TV and audio

If you have a standard television aerial, a satellite dish and maybe an FM aerial as well, you can combine the signals to run just one cable from the roof to the living room and then split them again to route to individual television and aerial components.

1 Fit a TV/FM diplexer, or triplexer if you have three signals to combine, in the loft or wherever your television and FM aerials are mounted. This simple socket allows two or three cables to be fed in and one single cable to run out, carrying multiple signals at slightly different frequencies.

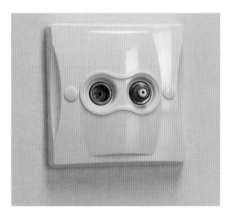

2 Run this cable to the point where you need the signals to diverge once more and fit a TV/FM diplexed twin socket (or a triple, if required) in the room.

BOOSTING THE SIGNAL

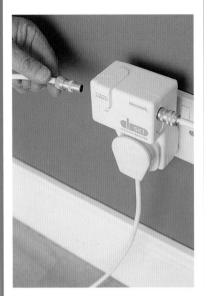

If the television signal where you live is weak, a simple plug-in amplifier can make a big difference to your picture quality. Plug the amplifier into the socket that powers your television then connect the lead from the aerial and the lead going out to the television.

Every time you split a signal to supply an additional television set, the signal to all the sets is weakened. Using a combined splitter and amplifier unit solves this problem, by boosting the signal where it first enters the house. Plug in the main 'in' cable from the aerial and take out separate feeds for each set in the house.

Installing your own telephone extensions

You can install telephone extensions easily by using kits sold in electrical shops, DIY centres and phone shops.

The law requires a master socket to be fitted by the company that provides the service. You should already have one; it is a square white socket. An old-style box will not accept the new plugs.

• Once you have a modern master socket you can install extensions throughout the house, provided there is no more than 100m of cable between the master socket, and the farthest extension socket, and no more than 50m between the master socket and the first extension socket.

• Extensions should not be fitted in damp areas such as bathrooms or toilets, or close to swimming pools.

• For safety reasons, telephone wiring must be kept at least 50mm from mains electric cables. The wiring works on a safe voltage, but it is advisable not to plug it into the master socket while doing the installation. If you need to join lengths of cable, use a telephone joint box (see page 111).

These instructions use British Telecom fittings; others may vary slightly.

Tools *Wire cutters; electrical screwdriver; trimming knife; bradawl or pencil; drill and masonry bit to fit wallplug provided with kit; screwdriver; small hammer; insertion tool (provided with kit).*

Materials *Telephone socket for each extension; socket converter attached to cable; cable clips; cleats; two or more telephone sets fitted with plugs. Possibly extra cable (without converter) and clips, depending on the number of extensions.*

Main telephone and extensions

The square white master socket must be installed by the company that provides the telephone service. This socket is the starting point for all extensions, which can be fitted anywhere in the house. A converter plugs into the master socket, and its cable runs to the first extension socket where the other end is wired in. Another cable runs to the next extension. Telephones are plugged into the converter and the sockets.

Wiring one extension

1 Decide where the extension socket is to be placed, and plan a route for the cable. It can be fitted along walls or skirting boards, or can run above ceilings and under floors.

2 Run the cable from the master socket to the point where the extension socket will be fitted. The converter on the cable goes at the master socket end, but do not connect it yet.

3 Fix the cable by hammering in the cable clips every 300mm, but leave the last few clips at the extension end until later.

4 Cut the cable where it reaches the extension-socket position, leaving about 75mm to spare.

5 Unscrew the front plate from the extension socket and use a trimming knife to cut away the entrance hole which is marked on the inside of the plate at the bottom.

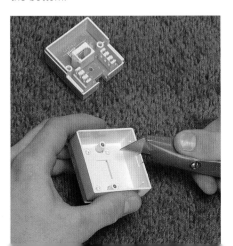

6 Using the trimming knife, strip about 30mm from the cable's outer sheath.

7 When there are six conductors, use the insertion tool to connect them to the socket, pushing them firmly into the grooves. Hold the tools as shown top right and push the green conductor with white rings into the connection marked 1; blue with white rings into connection 2; orange with white rings into connection 3.

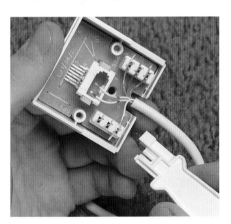

Now rotate the tool to push in the others, but be careful; the numbers 4, 5 and 6 run from bottom to top. Push the white conductor with orange rings into connection 4, white with blue rings into connection 5, and white with green rings into connection 6.

HELPFUL TIP

You will probably find it easier to wire the socket before fixing it to the wall.

8 Hold the back plate on the wall in the chosen position and mark the wall through the middle hole with a bradawl or pencil. Drill a hole in the wall to the depth of the wallplug. Insert the wallplug and screw the box to the wall.

9 Fit the front plate onto the back plate with the two screws.

10 Plug the converter into the master socket.

11 Fit the plug from the master telephone into the socket on the face of the converter.

12 Plug the extension telephone into its own socket.

Fewer than six conductors

If there are fewer than six conductors, leave connectors 1 and 6 empty or choose a four-connector extension socket.

Wiring extra extensions

You can install a second extension socket by running another cable from the first extension socket. No converter is needed. A third extension socket can be installed by running cable from the second, and so on.

1 Unplug the converter from the master socket and unscrew the front plate from the first extension socket.

2 Connect the conductors of the new cable in the same order as previously, on top of the existing ones. This can be done with the socket on the wall or, if you prefer, the socket can be temporarily removed.

3 Wire the other end of the cable to the second extension socket. Continue in this way for any further extensions. Fit the new conductors on top of those already there.

4 Screw the front plates on the sockets and plug in the converter and the telephones.

What if it fails to work?

If a newly installed telephone fails to work, carry out this fault-finding procedure:
• Unplug the converter from the master socket, plug in the telephone and make a call. Test each extension telephone by plugging it into the master socket as well.

• If none of them works, the fault is in the wiring installed by the telephone company.
• If one of the phones fails to work, it is probably faulty. Return it to the supplier.
• If they all work, the fault lies in your new wiring, so check each socket to make sure that you have wired the coloured conductors to the correct numbers and pushed them right in. If the conductors are wrongly connected, pull them out one at a time with long-nosed pliers, cut off the used core and re-make the connection.
• If the sockets are properly connected, check that the cable is not kinked, cut or squashed. If it is, replace it.
• If the phones still do not work, return the sockets, joint box converter and cable to the suppliers and ask for a replacement.

More than one extension from a single point

There are some situations where you may want to run more than one extension from a single point. In this case, run cable from the converter at the master socket to a telephone joint box which will take that cable plus as many as three extensions.

1 Fit cable between the master socket and the position for the joint box. Do not plug in the converter yet.

2 Unscrew the front plate from the joint box and cut out the plastic for the cable at the top of both plates.

3 Fit the conductors into the joint box in the same way and the same order as in an ordinary socket. The one difference is that the six connectors are in two linked lines down the sides rather than to the left and right, so strip off about 40mm of the cable sheath to fix them. Fit one of the extension cables.

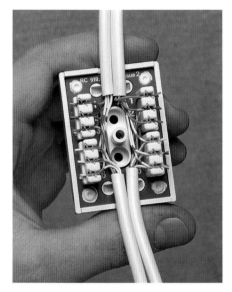

4 Fit the remaining extension cables in the same way, feeding them in from the top or the bottom.

5 Push the cable straps under the pairs of cables and through the holes at the top and bottom of the back plate, and pull them tight. Then cut off the tails.

6 Fix the backplate to the wall with the two wallplugs and screws provided.

7 Fit the extension cables to their sockets in the normal way.

8 Screw the front plates on all the sockets, then plug in the converter and all the new telephones.

TELEPHONE WIRING COLOURS

You may find different core insulation colours in a telephone extension kit – for example, a typical kit has cable with green, yellow, orange and black. Make sure you follow the instructions supplied with the kit to connect the wires to their correct terminals in the extension socket.

COMPONENTS FOR DIY TELEPHONE WIRING

Extension sockets are smaller than the master socket that is installed by the telephone company. They are wired up with a plastic insertion tool supplied with the extension kit. A converter is supplied already attached to a long length of cable that is fixed in place with cable clips. A telephone joint box can be used to take up to three extensions from one point, as well as the cable. It can also be used to join two lengths of cable together if necessary.

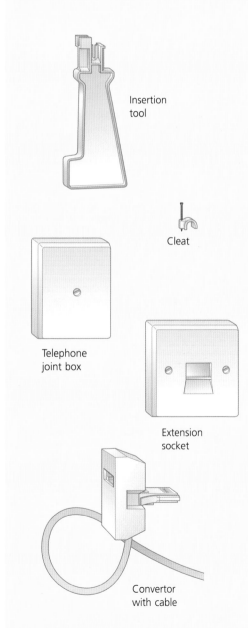

Insertion tool

Cleat

Telephone joint box

Extension socket

Convertor with cable

Wiring options for home computers

Many homes now routinely have more than one computer. Linking these together as a network allows the individual workstations to share a printer or internet connection or to transfer files from one to another automatically.

Benefits of a home network
Linking all the computers in the house on a single network makes it easy to transfer files from one to another. If you have one computer with a significantly larger hard drive than the others, you can also use that computer as a file server for the additional machines or as a back-up for the information stored on the individual PCs.

Another major benefit of a home network is the ability to share peripherals – such as printers, scanners or Webcams – and to access the internet from any of the computers on the network via a single connection. Having just one computer 'live' online is a more secure option than having several computers simultaneously connected to the internet. With a high-speed broadband connection, more than one computer can be connected and online at any one time – or permanently.

Home network options
A home network system or LAN (Local Area Network) brings together all the systems that transmit electronic information around the house. This means that it is possible to integrate other services and systems into the network such as cable television, home entertainment systems, security systems and phone lines. The older traditional method is the peer-to-peer wiring method, where various jack sockets or connectors are installed along a single loop running throughout the house – in much the same way as telephone extensions are wired. Although simple to install, this kind of primitive network cannot handle voice and data signals at the same time. In addition, the more devices such as computers that are placed on the line, the more the network signal is weakened.

Wired or wireless? A hard-wired network of Ethernet cables linking computer to computer is the cheapest and most reliable option, whichever network design you choose to use (below). Transfer of information is often faster than with a wireless signal, but the initial set-up will cause greater disruption to the house and you will have to find ways of concealing cables or making them safe. A wireless network does away with trailing cables and fixed workstations. It offers flexibility that a wired network cannot match and is particularly well-suited to lap-top computer users. However, the hardware is more expensive than that required for a wired system and the connection will be less reliable, and often also slower.

Peer-to-peer networks The very simplest and oldest type of network is a loop-in system (above) where computers are linked in a string, known as a peer-to-peer system. In a home with only two computers, the workstations can be fitted with a network card and wired together with a single Ethernet cable. This system can also be used for more computers, but the quality of the signal is likely to be poor and if a problem develops with just one of the outlets in the string, the entire network will be affected.

Star networks Newer network systems resolve these problems by using what is called star topology in which all the cables are distributed

Connecting computers in a star network

Internet connection — Printer — Hub — Ethernet cable — Networked computer from a central point. With this arrangement, if one connection fails it is unlikely to affect all the rest of the computers or other devices on the system. The ideal is to install all the cables when a house is built or replastered, burying them beneath floorboards or within the wall and terminating in sockets around the house. However, it is also possible to install these in existing properties, running cables beneath floorboards and above ceilings.

The central distribution point of the network may be a hub, wired Ethernet switch or wireless router, depending on your requirements. A hub (below left) will connect up to four computers in an average home network reliably and cost-effectively. If you have more than four computers to link, or if you often transfer very large files, such as music or video files or want the facility for high-speed online gaming, then a wired switch may be a faster and more reliable option.

CAT5 wiring If you are building a new house or completely renovating an existing property, it is worth considering installing CAT5 data cabling throughout. This is a more powerful Ethernet cable than the standard linking cable used in most wired networks, and is capable of carrying more data, much faster.

Even if you currently only use your home network for computer file-sharing, fitting a CAT5 network will give you the options of running telephone, computer, music or audiovisual signals throughout the house at a later date. The network also has the potential for upgrading to more sophisticated household electronics, such as room-to-room intercoms or door entry systems, touch-sensitive heating and lighting controls in each room, internet access from terminals other than home computers and much more.

Wireless networks A wireless, or 'WiFi' network works from a hub called a router, which is wired into the internet connection, if you have one, and communicates with the computers on the network via a radio signal. Each computer is fitted with a Network Interface card: a credit card sized transmitter, and receiver that communicates with the base station, although most wireless routers allow for a combination of wired and wireless connections with computers on the network. Either the router or one of the terminals on the network may be connected to peripheral devices, such as a scanner, that each of the network computers can access.

Most wireless routers have a range of up to 45m, but this is reduced if the signal has to pass through walls, particularly solid brick walls. To get the best connection to all the computers on the network, position the router high up in the house and as central as possible. A badly placed base station will lead to the signal dropping off and one or more of the computers becoming disconnected. Follow the instructions that come with the router for more specific advice on positioning.

Wireless routers are prone to interference from other household devices, such as microwave ovens, cordless telephones and automatic garage door openers. Consider this when deciding where to position your base station – putting it too close to the main cordless telephone in the house may affect the reliability of the network signal.

WHAT IS ADSL?

ADSL stands for asymmetric digital subscriber line. What this means is that digital signals are sent over phone lines to and from your home computer. Even though you may only have one line into your home the ADSL recognises the difference between voice and data so you are able to be online and talk on the telephone at the same time. Using an ADSL connection, files downloaded from the internet are faster than those that are uploaded, in other words sent. If you will be sending files regularly then consider investing in an SDSL connection, or symmetric digital subscriber line. With this, data can be sent with the same speed in either direction.

SHARING AN INTERNET CONNECTION

For several networked computers to access the internet through a single dial-up connection, the PC fitted with the modem must have enabled Internet Connection Sharing. Broadband routers make the connection themselves, eliminating the need for a designated internet 'host' computer. They also have the added benefit that they often come with built-in firewall support.

Installing a home computer network

Installing a home network allows you to link several computers together to access a single internet connection, transfer or back up files, or share peripheral devices, such as a printer or scanner.

Tools *Drill; screwdriver; possibly also wire strippers and trimming knife.*

Materials *Network router; required lengths of Ethernet cable; required length of telephone or modem cable; broadband microfilter, if appropriate; possibly lengths of plastic trunking to conceal cables.*

1 Mount the network router on the wall in a location close to the incoming internet cable (telephone socket) and to a mains socket outlet. Follow the manufacturer's instructions: you may need to fit a wall bracket, or simply fix two screws protruding from the wall, on which the router hangs.

2 Decide on the route the network cables will take to reach the intended computers. Either prepare the route for burying the cables in the wall (page 57) or fit lengths of plastic trunking along the skirting boards and door architraves. Connect the modem cable from the telephone socket to the router (above), running via a broadband microfilter if you use an ADSL internet connection (page 113).

3 Run Ethernet cables out from the router to the computers it will serve. Tack the cables to the skirting board, bury them in the wall, or conceal them in plastic trunking to keep them out of the way.

4 Plug in the router to the socket outlet and follow the manufacturer's set-up instructions to establish the connection for each computer on the network.

Choosing a burglar alarm

Most thieves are likely to be deterred by locks on windows and doors, but you may decide to install an alarm system as an extra defence against a burglar who tries to force his way in. A noisy alarm may deter him from entering the house, or greatly reduce the time he stays there.

Before buying a system, check that the alarm is loud enough. Anything below 95 decibels has little effect and cannot be heard over any distance. Most alarms sound for about 20 minutes and then reset to avoid nuisance to neighbours. Notify your local police and your neighbours that an alarm has been fitted, and give a trusted neighbour a spare key to the system or the code number. A valuable addition to a system is a panic button, which can be used to trigger the alarm at any time. Some burglar-alarm systems are designed for DIY installation; others need to be professionally fitted.

WHOLE-HOUSE SYSTEMS THAT WARN OF A BREAK-IN

The most common alarm systems are designed to set off a bell or siren if a burglar tries to break in. Before buying one, ensure that the alarm has a 'closed' electrical circuit. This means that when the system is turned on the circuit is completed. If there is any interference – such as the wires being cut – the alarm will go off. Each system has three main components – the switches and detectors, the control unit and the alarm.

Control unit The 'brain' of the system is the control unit, which receives signals from the switches and sends an electric current to activate the alarm. The system is turned on or off with a key or a push-button panel to which a code number is first keyed in.

Connected with the control unit will be some form of power supply, either mains or battery. In most systems, mains power will feed the system under normal conditions, but if the power is cut off for any reason, a battery will take over. The battery is automatically recharged when power is restored.

Passive infra-red (PIR) motion detectors

These are small units, fitted at ceiling height, which sense movement through changes of temperature within their field of detection.

Panic button
A manually operated switch can be fitted as a panic button – at the bedside or by the front door. It is usually wired so that it will trigger the alarm whether or not the rest of the system is switched on. Panic buttons can be very sensitive; the slightest pressure will set off the alarm.

External alarm The alarm has a bell or siren that should be loud enough to frighten off potential intruders and alert neighbours. Some external sirens also incorporate a bright, flashing strobe light; others feature flashing LEDs to indicate that the system is active and to add to the visual deterrent. The external alarm box should be tamper-proof and should contain its own battery, so that it will still sound if the cable to it is cut.

Magnetic door and window contacts A magnetic switch can be fitted to a door or window that opens. One part is secured to the frame, the other to the door or casement. If the magnet is moved the circuit is broken and the alarm is triggered.

Door alarm This unit can be fitted to an external door and will set off an alarm if the door is opened. It is battery operated, so no wiring is needed. The alarm is turned off with a push-button code that you set yourself, or with a key. A delay switch allows several seconds for you to enter or leave the house without triggering the alarm.

Shed alarm This type of self-contained battery-powered alarm can be fixed in a shed, garage, caravan or greenhouse. It will pick up movement within its field of detection and activate its own alarm.

Wireless alarm systems Passive infra-red sensors and magnetic door contacts are independently battery powered and transmit a radio signal to the control unit when they sense movement; this in turn triggers the alarm. These 'wirefree' systems can be set using a remote key fob switch which doubles as a mobile panic alarm. Because the kits remove the need to run cables, they are much quicker to install. Some systems offer a repeater unit which increases the transmission range so that outbuildings may also be protected.

Closed-circuit television (CCTV) systems CCTV systems are now within the budget of most ordinary householders. Cameras are compact and systems that you can install yourself are widely available. You can choose whether to have the camera tied to its own dedicated monitor, or to utilise your own TV and video set-up. Some cameras incorporate passive infra-red (PIR) detectors to start recording only when movement is detected. When used to see who is at the front door, the sensor will automatically switch from the channel you are watching, to the surveillance camera.

Choosing time controls for security

A dark, silent house can arouse the interest of burglars. If you make the house appear occupied, by day or night, a prowler is likely to move on to an easier target.

Voices, music and lights after dark can suggest you are in – but only if used with discretion; a single light in the hall left on all evening, or a radio playing all day are more likely to betray that you are out. The illusion that you are there is given by a change in the house – music stopping, or a light going off in one room and on in another. Sockets and switches operated by timers help to create the illusion. Some sophisticated controls will memorise your schedule of switching lights on and off and reproduce it.

Light-switch timers
Time-controlled light switches are connected to the lighting circuit in place of normal switches and are wired in the same way (page 85). Some will switch on and off many times. Others switch on only once – at dusk – so there is no danger of the light shining in the daytime because a power cut has interfered with the setting.

Sensor lamp and light sensor switch
Sensors automatically switch a low-energy bulb on and off at sunset and sunrise; the lamp has a 10,000 hour life. Alternatively, a light sensor switch can be used with tungsten lighting, or fluorescent or low-energy fittings. The switch is controlled by a light-sensitive photocell that activates it when daylight fades. The switch-off time is set at 1 to 8 hours after the light comes on.

Automatic outdoor light
An infra-red sensor activates the light when anyone – visitor or burglar – comes within its field of vision. The sensor can be incorporated in either the light or a separate unit operating a number of lights. Once on, the light shines for a period chosen by the user.

Programmable switch
Digitally controlled, these switches offer a sophisticated combination of on/off switching and are programmable to come on at different

times during the week. Features may
include light sensor settings and an override. Most can only be used with tungsten
lighting, but others are adapted to use low-
voltage or fluorescent lighting.

Plug-in timers A plug-in timer will control
any appliance that plugs into a 13amp
socket outlet – a radio or a lamp, to give
the impression that the house is occupied,
or a heater or electric blanket for
convenience. Set the required programme
on the timer, plug it into a switched-on
outlet, plug the appliance into the timer
and switch on the appliance. It will not
actually come on until the programmed
time. You can override the set programme
manually. There are 24-hour and seven-day
timers. They vary in the number of times
they will switch on and off, and in the
shortest possible 'on' period.

24-hour timer Some
plug-in timers allow
you to design a
pattern of switching
on and off over a
24 hour period.
Some models allow
up to 48 changes in

a day. Markers on a dial trigger the timer to
switch on or off as the dial turns. The
shortest 'on' period is usually 15 to 30
minutes. The pattern will be repeated every
day until it is switched off or the markers
are changed.

Seven-day timer
These work in the
same way as 24-
hour timers, but
switching patterns
can be varied from
day to day over a

period of seven days; some will allow up to
84 switchings per week. The on/off pattern
will then be repeated each week until
altered. The minimum period for which it
can be switched on is two hours.

Electronic digital timer
These timers offer a
number of setting
programmes and may
allow up to 84
switchings a week. A
random setting will
switch lights or a radio

on or off during the
power-on period, to give the impression
that a house is occupied. It can be used as
a daily or weekly timer and will
automatically repeat programmes.

PROTECTION FROM FIRE: SMOKE ALARMS

Building Regulations require mains-
powered smoke alarms to be fitted in
new properties and existing homes that
are extended or where the loft has
been converted. The Government
recommends that at least two alarms
approved to British Standard BS5446
Part 1 are fitted in an average two-
storey house – one downstairs in the
hall and the other on the landing.

Smoke alarms should be fitted
within 7.5m of the door to every
habitable room – living rooms, kitchens
and bedrooms. An alarm fitted to the
ceiling should be at least 300mm
away from a wall or ceiling light
fitting. Some alarms have an
escape light fitted; some
can be linked – so that
if one detects smoke,
all are activated.

Smoke alarms for
existing homes can be
battery-operated, wired to
the mains with a back-up
battery or plugged into a ceiling light
fitting (below).

There are two basic types, ionisation
and photoelectric. For good, all round
protection use one of each.
• The ionisation alarm works by
detecting invisible smoke particles in the
air and responds quickly to fast, flaming
fires, so it is a good choice for a
bedroom, hall or landing. It is not
suitable in or near a kitchen because it
will be set off by cooking fumes.
• The photoelectric alarm 'sees' smoke
and is more sensitive to smouldering
fires which usually occur in furniture.

Some battery-powered alarms have
indicators to show that the battery is
still working; others give an audible
warning when the battery is low. Plug-
in alarms recharge when the light is
switched on. Accumulated dust impairs
performance, so clean alarms regularly
using a vacuum cleaner nozzle.

Wiring to an outbuilding

An electricity supply to a detached garage, greenhouse, garden shed or other outbuilding must be run as a permanently fixed cable from the consumer unit in the house. A long flex run to the outbuilding from a socket inside the house is not acceptable.

The circuit to an outbuilding can be run overhead or underground.
• It is best to bury the cable despite the work involved. Cable underground does not mar the view, and with careful planning the length of the underground section can be kept to a minimum. Run the cable through the house to reach the point nearest the outbuilding to save work. If this is difficult, run the cable along the outside wall of the house to reach the point nearest the outbuilding. You can also run it along a boundary wall, but not along a fence.
• You can take the cable overhead, but it is unsightly to have it hanging immediately outside the house. It should be fixed at

Cables and fittings for outdoor wiring

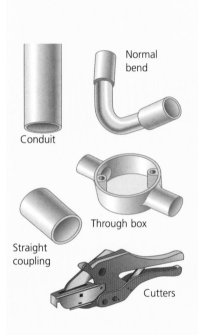

Conduit
Normal bend
Straight coupling
Through box
Cutters

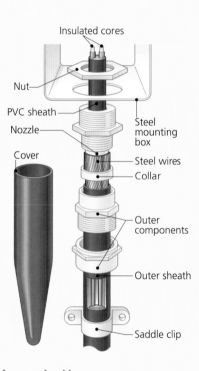

Insulated cores
Nut
PVC sheath
Nozzle
Cover
Steel mounting box
Steel wires
Collar
Outer components
Outer sheath
Saddle clip

Plastic conduit
Plastic (PVC) conduit can be used for protecting cable run in solid walls, to protect surface wiring run in an outbuilding and for underground wiring as described on these two pages. It comes in black and white, in various sizes (from 16mm to 50mm) and in two grades – make sure you use the high-impact (heavy gauge) grade underground. You can get straight couplers, curved connectors (bends and elbows), T-connectors and a whole range of special boxes and support clips. The conduit itself comes in 3m lengths, but can easily be cut with a hacksaw or special cutters; joining to fittings is done by solvent welding.

Armoured cable
Two insulated conductors are in a PVC sheath inside steel wires that serve as the earth conductor. There is a PVC outer sheath. Three-core armoured cable has three insulated cores, coloured brown, grey and black – the grey core can be used for the earth covered at the ends with green-and-yellow sleeving. Armoured cable needs no further mechanical protection, making it ideal for use outside and in outbuildings.

Use the right size gland to seal the cable into a deep steel mounting box. On the two-core cable, tighten the gland nut hard to complete the earth continuity. Use plastic, galvanised or copper clips to hold the cable.

least 3.5m above the ground, which frequently entails fitting a timber post securely to the outbuilding to achieve the clearance. If it is more than 3m long, the cable span must be supported along its length by a metal wire (called a catenary) and must have a drip loop at each end.
• For underground cable, make the route of the trench as short as possible, well clear of obstacles such as rockeries, inspection chambers or trees. Allow some extra cable when you calculate the length needed in case you meet unexpected obstacles.

A professional electrician is likely to install armoured cable (see Cables and fittings for outdoor wiring, left), but this is fairly expensive and needs special connecting glands, which require some skill to use. It is easier to use ordinary two-core-and-earth cable for the entire sub-circuit between house and outbuilding, but the outdoor section of the circuit cable must be run in protective PVC conduit fixed to the walls of the buildings and laid in the trench. The run is made up using lengths of conduit linked with solvent-welded straight and elbow connectors.
• A 20A radial circuit can be run from a spare RCD-protected 20amp MCB in the consumer unit, but if this is not available, fit a new enclosure as described on page 55 with a 63amp/30mA RCD and one or two MCBs.
• Call in your electricity supply company to connect a new unit to the meter once you have installed the new circuit wiring. You are not allowed to do this yourself.
• Inside the outbuilding, the cable should run to a socket outlet. A switched fused connection unit (FCU) can be run from this to provide a starting point for a fused lighting sub-circuit.

Tools *Tape measure; side cutters; wire strippers; power drill; long 20mm masonry bit; hammer; spade; hacksaw; trimming knife; screwdrivers.*

Materials *2.5mm² two-core-and-earth cable; cable clips; green-and-yellow PVC earth sleeving; brown PVC insulating tape; 20mm PVC conduit; wall clips; straight connectors and elbows; solvent-weld adhesive; sand; bricks; paving slabs; warning tape; enclosure with 30mA RCD and one or two MCBs; metal-clad double socket outlet and mounting box; metal-clad FCU and mounting box; 1mm² two-core-and-earth cable; batten lampholder; metal-clad light switch and mounting box; grommets; wood-screws; wallplugs.*

Work inside the house

1 Plan and prepare the cable route from the consumer unit or new enclosure to the point where it will leave the house. With suspended timber floors, run it beneath the floorboards.

2 Where the circuit cable will leave the house, locate a mortar course in the wall at least 150mm above the level of the damp-proof course. Use a long 20mm diameter masonry drill bit to drill from the outside, sloping the hole slightly upwards to discourage rainwater from entering and mortar in a short length of conduit.

3 Mount the new RCD unit on a board close to the meter. Remove knockouts from the base of the unit to admit the incoming meter tails and circuit cable. Clip the RCD and the MCBs onto the metal busbar, ready for connection later.

Outside work

1 Plan the cable route between the house and the outbuilding, and excavate the trench along the marked route. Avoid cultivated ground if possible; a route running close to a path or boundary wall is often the ideal choice.

2 Check the depth of the trench. In open ground, it should be at least 500mm deep.

3 Remove any sharp stones from the trench and put in a layer of dry sand about 50mm deep to protect the conduit.

4 Measure the cable run and work out how many lengths of conduit you will need. Conduit is available in 3m lengths.

5 At the house end of the run, feed the end of the circuit cable into the conduit in the wall. Pull enough cable through the conduit to make up the run inside the house from the entry point in the wall to the meter position and then unroll enough cable to complete the run.

6 Use a short length of conduit with an elbow solvent welded on one end to connect to the conduit in the house wall and a large-radius elbow to connect to the conduit underground. Clip this to the wall. Assemble the conduit run by solvent-welding lengths of conduit together with straight joints. Brush a little special adhesive onto the pipe end – do not get any on the cable sheath. Push the fitting onto the pipe and rotate it slightly to ensure a waterproof joint. Feed the whole cable run through the next length of conduit before joining it to the previous one. Add lengths of conduit one by one in the same way.

7 Use a hacksaw to cut the last section to length as necessary. Smooth off any roughness using fine abrasive paper before joining it to the previous length of conduit.

8 When you have joined all the lengths of conduit together, lower the completed run into the base of the trench. Use a large-radius bend and a short length of conduit to take the run up to the point where it leaves the house and where it enters the outbuilding.

WEATHERPROOF SOCKETS

Sockets and light switches that are fitted out of doors must be weatherproof. Socket outlets on the house wall can be wired as spurs from an indoor power circuit (see page 63), but must be protected by a high sensitivity (30mA) RCD (see page 122).

9 Pass a short length of conduit through the outbuilding wall and connect it to the underground run with a standard elbow. Secure exposed conduit with clips. Feed enough cable into the outbuilding to allow it to be connected to the wiring accessories that will complete the sub-circuit.

10 When you have completed the conduit run and laid it in the trench, place a line of bricks on the sand bed at each side of the conduit and lay narrow paving slabs over the bricks. This ensures that any future digging cannot damage the conduit or the cable inside it. Lay lengths of special yellow-and-black plastic warning tape over the slabs.

11 Back-fill the trench with the soil you excavated earlier, and tamp it down firmly.

Work in the outbuilding

1 Decide where in the outbuilding you want to position the various wiring accessories. Because the sub-circuit is controlled and protected by the RCD and MCB at the house end of the circuit, the simplest way of wiring up the accessories is to create a radial circuit. Run the incoming cable to the first socket, then on to any other sockets you want to install. At a convenient point in this radial circuit, include a fused connection unit (FCU) fitted with a 3amp fuse. From here a length of 1mm² cable is run to supply one or more light fittings and a separate light switch.

2 Screw the metal mounting boxes for the various wall-mounted accessories to the wall of the outbuilding. Use woodscrews driven direct into timber walls or frame members, and into wallplugs in drilled holes in masonry walls.

3 To protect the cable within the outbuilding, fit plastic conduit, supported by saddles, between the mounting boxes. The conduit is connected to the boxes by glands that fit into the box knock-out holes.

4 Feed the incoming cable into the first mounting box and cut it to leave about 150mm of cable within the box. Then run cable from this box to the next one, cutting it to length as before. Carry on adding lengths of cable one by one to connect all the wiring accessories. If you have difficulty pushing the cable through the PVC conduit, you can get draw tape to help.

5 Prepare all the cable ends for connection by stripping about 75mm of insulation from each conductor. Cover the bare earth cores with lengths of green-and–yellow PVC sleeving.

6 There will be two sets of conductors to connect to the socket faceplate at each socket (except at the last one on the circuit, which will have just one set). Connect the two brown conductors to the terminal marked L and the two blue conductors to the terminal marked N. Connect the sleeved earth conductors to the earth terminal on the faceplate. Add a short length of sleeved earth core (taken from a cable offcut) between this terminal and the earth terminal in the mounting box to earth the mounting box.

7 Check that all the connections are secure. Fold the conductors back into the mounting box and screw the faceplate to it. Repeat this process to connect up the other sockets on the sub-circuit.

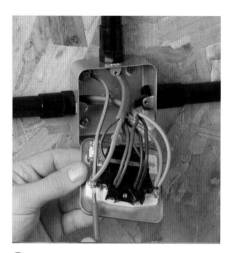

8 At the fused connection unit (FCU), connect the conductors of the incoming cable to the terminals marked FEED or IN. The brown conductor goes to the terminal marked L, the blue conductor to the terminal marked N and the sleeved earth conductor to E or ⏚. Add an earth link between the earth terminals on the faceplate and the mounting box as before. Connect the 1mm² cable that will supply the light fitting in the same way, but to the terminals marked LOAD or OUT. Run this cable up to the position of the batten holder or light fitting you are installing. Fit a 3amp fuse in the fuseholder and screw the faceplate to the mounting box.

9 At the batten holder or light fitting, connect in the 1mm² cable from the FCU and another 1mm² cable to run onto the light switch. The two brown conductors go to the centre bank of terminals, and the blue conductors to the end terminals; it does not matter which way round these two are connected. Connect the two earth conductors to the earth terminal.
Screw the cover on to the batten holder.

Connecting at the RCD unit

At the new MCB/RCD unit in the house, connect the brown conductor of the circuit cable to the top terminal of one of the MCBs. Connect the neutral and earth cores to their respective terminal blocks.

The second MCB shown here is wired to supply another sub-circuit. Your electricity supply company will fit the RCD meter tails and add the main earth cable.

Fitting an outdoor socket

Using power tools in the garden doesn't have to mean trailing extension leads out of a window. Run a spur from an indoor socket outlet circuit and install a weatherproof outdoor socket. If the circuit is not already RCD protected, use an outdoor socket with its own RCD.

1 A neat way to fit an outdoor socket is to drill a hole behind an existing indoor socket outlet on an outside wall (lined up with one of the mounting box knock-outs) and locate the new outdoor socket on the other side of the wall.

SAFETY TIP

Test the RCD before you start work every time you use your outdoor outlet by pressing the TEST button. If the RCD trips off, press RESET: the outlet is ready to use. If it does not, there is a fault, and you should call an electrician.

2 Turn off the mains power and remove the faceplate and mounting box of the socket you will be working from. Drill a pilot hole then switch to a 10mm masonry bit at least 300mm long to drill through the wall. With the power switched off, you will need a cordless drill with a hammer action, to get through the wall.

5 Cover the bare earth core with PVC sleeving and connect the cores to the terminals on the faceplate. There is no need to earth a plastic mounting box. Screw the faceplate to the box. Restore the power and test both sockets before use.

Fitting an outdoor wall light

To install a light on an outside wall– under a porch or over a patio, follow the same method as for an inside wall light. Choose a weatherproof light fitting designed for use outside.

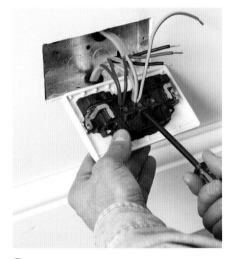

3 Screw back the mounting box and feed the new cable through the wall. Connect the old and new cores to the faceplate. There will be three cores in each terminal: the ring main cores coming in and going out, and the new spur. Reconnect the 'flying earth' cable to the box earth terminal and replace the faceplate.

4 For safety, use an outside socket with an entry point through the back. Any exposed cable on an outdoor wall should be armoured or protected in a conduit (see page 120). Use the grommet supplied to make the cable hole watertight, drill out any drainage holes indicated and screw the box to the wall, after first running some silicon sealant round the box/wall junction.

If the light is on a bracket, not a flush-fitting baseplate, you will need waterproof sealant or a rubber gasket to fit the bracket's base.

It is safest to have the switch indoors, though you may want to have an outdoor switch as well. You can take the cables for the light and the switch along the same route over the ceiling and down the indoor surface of the external wall on which you are going to install the light.

At the right height for the wall fitting – about 1.8m above ground level – drill from the back of the chase through the inner and outer leaves of the external wall to make a hole to lead the cable outside. Feed the light cable out through the hole.

Install and connect the junction boxes and the switch in the same way as for an indoor wall light (see page 91).

● Cable core colours have changed (see page 56)

Wiring to an outdoor wall light

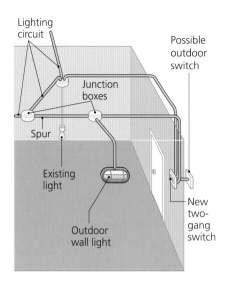

Lighting circuit

Possible outdoor switch

Junction boxes

Spur

Existing light

Outdoor wall light

New two-gang switch

Light on a base plate

1 Hold the baseplate of the light fitting against the wall and probe with a bradawl through the fixing holes to mark drilling spots on the wall. Drill holes at the marks and insert wallplugs in them.

2 Feed the cable through the baseplate and screw the plate to the wall. Prepare the cable for connection (see page 39).

3 Connect the brown and blue cores to the terminals of the lampholder; it does not matter which core goes to which terminal. Connect the green-and-yellow sleeved earth core to the terminal marked E or ⏚.

Light on a bracket

1 If your light is on a bracket, not a baseplate, you cannot reach the lampholder. Flex will have been connected to it already and the flex end will stick out from the bracket. Prepare the flex cores and connect them to a terminal connector block.

At the terminals on opposite sides of the block, connect the cores of the light cable. Link brown to brown, blue to blue and sleeved earth to earth.

Coat the rim of the bracket's baseplate with sealant, or fit a rubber gasket.

2 Screw the light fitting into the drilled and plugged holes. Make sure that any sealant or gasket is squeezed tight against the wall to make a weatherproof seal.

Fitting an outside switch

If you want a switch out of doors as well as inside for the new light, use a two-way sealed splashproof design. Put it on the outside wall back to back with a two-way indoor switch. You will need an additional short length of three-core-and-earth 1mm² lighting cable for the connection between the two switches. Drill a hole between the switches and feed the three-core-and-earth cable through it. Connect the cable as described on page 85.

Acknowledgments

All images in this book are copyright of the Reader's Digest Association, Inc., with the exception of those in the following list.

The position of photographs and illustrations on each page is indicated by letters after the page number:
T = Top; **B** = Bottom; **L** = Left; **R** = Right; **C** = Centre
Front cover Fotalia/Borodaev; istockphoto/James Weston; **39 L, TR** GE Fabbri Limited; **40 L, R** GE Fabbri Limited; **41 BL** GE Fabbri Limited; **42 TL, TR, BL, BR** www.screwfix.com; **49 BC** HL Studios, Oxford; **87 R** GE Fabbri Limited; **90 TR, L, R** GE Fabbri Limited; **91 TL** GE Fabbri Limited; **95 TL** www.screwfix.com, **BR** Electric Light Company; **96 TL, TR** www.ring.ltd.uk, **BR** GE Fabbri Limited; **97 TL, TR, BR** GE Fabbri Limited; **99** GE Fabbri Limited; **100 TL, TR, BR** GE Fabbri Limited; **101 TL, TR, C** GE Fabbri Limited; **108 TL** GE Fabbri Limited; **116 CL, CR, BR** www.screwfix.com; **117 TR** Cooper Security Ltd, **TL, CL, BL** www.screwfix.com; **CR** www.fireangel.co.uk.
The Reader's Digest Association Inc. would like to thank Draper tools (www.drapertools.com) for the loan of tools, props and other materials.

Reader's Digest DIY Wiring and Lighting
This edition published in 2012 in the United Kingdom by Vivat Direct Limited (t/a Reader's Digest), 157 Edgware Road, London W2 2HR

First published as **Reader's Digest Wiring and Lighting Manual** in 2005

Project Editor Jo Bourne

Art Editor Sailesh Patel

Consultant David Holloway

Editorial Director Julian Browne

Art Director Anne-Marie Bulat

Managing Editor Nina Hathway

Trade Books Editor Penny Craig

Picture Resource Manager
  Sarah Stewart-Richardson

Pre-press Account Manager Dean Russell

Production Controller Jan Bucil

While the creators of this work have made every effort to ensure safety and accuracy, the publishers cannot be held liable for injuries suffered or losses incurred as a result of following the instructions contained within this book. Readers should study the information carefully and make sure they understand it before undertaking any work. Always observe any warnings. Readers are also recommended to consult qualified professionals for advice.

Typesetting, illustration and photographic origination
Hardlines Limited, 17 Fenlock Court, Blenheim Office Park, Long Hanborough, Oxford OX29 8LN

Origination FMG
Printed and bound in China

Reader's Digest DIY Wiring and Lighting is based on material in **Reader's Digest DIY Manual** and **How Everything in the Home Works**, both published by The Reader's Digest Association, Inc.

Reader's Digest DIY Wiring and Lighting is owned under licence from the Reader's Digest Association, Inc. All rights reserved.

Copyright © 2012 The Reader's Digest Association, Inc.
Philippines Copyright © 2012 The Reader's Digest Association Far East Limited
Copyright © 2012 The Reader's Digest Association (Australia) Pty
Copyright © 2012 The Reader's Digest Association India Pvt Limited
Copyright © 2012 The Reader's Digest Association Asia Pvt Limited

We are committed both the quality of our products and the service we provide to our customers.
We value your comments, so please do contact us on **0871 351 1000**, or via our website at
www.readersdigest.co.uk

If you have any comments about the content of our books, email us at **gbeditorial@readersdigest.co.uk**

ISBN 978 1 78020 126 9
BOOK CODE 400-597 UP0000-1